Introduction to
Marine Engineering

Second Edition

Introduction to Marine Engineering

D. A. Taylor, MSc, BSc, CENG, FIMarE, FRINA
Senior Lecturer in Marine Technology, Hong Kong Polytechnic

Butterworths
London   Boston   Singapore   Sydney   Toronto   Wellington

PART OF REED INTERNATIONAL P.L.C.

First published 1983
  Reprinted 1985
Second edition 1990

© Butterworth & Co (Publishers) Ltd, 1990

British Library Cataloguing in Publication Data

Taylor, D.A. (David Albert), *1946–*
  Introduction to marine engineering. — 2nd ed.
  1. Marine engineering
  I. Title
  623.87

  ISBN 0-408-05706-8

Library of Congress Cataloging-in-Publication Data

Taylor, D. A., M.Sc.
  Introduction to marine engineering/D.A. Taylor. — 2nd ed.
    p. cm.
  ISBN 0-408-05706-8 :
  1. Marine engineering. 2. Marine machinery. I. Title.
VM600.T385 1990
623.87—dc20   89-71326

Phototypeset by Scribe Design, Gillingham, Kent
Printed in Great Britain at the University Press, Cambridge

Preface to second edition

Progress has been made in many areas of marine engineering since the first edition of this book was published. A greater emphasis is now being placed on the cost-effective operation of ships. This has meant more fuel-efficient engines, less time in port and the need for greater equipment reliability, fewer engineers and more use of automatically operated machinery.

The marine engineer is still, however, required to understand the working principles, construction and operation of all the machinery items in a ship. The need for correct and safe operating procedures is as great as ever. There is considerably more legislation which must be understood and complied with, for example in relation to the discharging of oil, sewage and even black smoke from the funnel.

The aim of this book is to simply explain the operation of all the ship's machinery to an Engineer Cadet or Junior Engineer who is embarking on a career at sea. The emphasis is always upon correct, safe operating procedures and practices at all times.

The content has been maintained at a level to cover the syllabuses of the Class 4 and Class 3 Engineer's Certificates of Competency and the first two years of the Engineer Cadet Training Scheme. Additional material is included to cover the Engineering knowledge syllabus of the Master's Certificate.

Anyone with an interest in ships' machinery or a professional involvement in the shipping business should find this book informative and useful.

D.A. Taylor

Acknowledgements

I would like to thank the many firms, organisations and individuals who have provided me with assistance and material during the writing of this book.

To my many colleagues and friends who have answered numerous queries and added their wealth of experience, I am most grateful.

The following firms have contributed various illustrations and information on their products, for which I thank them.

Aalborg Vaerft A/S
AFA Minerva
Alfa-Laval Ltd
Angus Fire Armour Ltd
Asea Brown Boveri Ltd
B & W Engineering
Babcock-Bristol Ltd
Babcock Power Ltd
Beaufort Air–Sea Equipment Ltd
Blohm and Voss AG
Brown Bros. & Co. Ltd
Caird & Rayner Ltd
Cammell Laird Shipbuilders
Chadburn Bloctube Ltd
Clarke Chapman Marine
Combustion Engineering Marine
  Power Systems
Comet Marine Pumps Ltd
Conoflow Europa BV
Deep Sea Seals Ltd
Doncasters Moorside Ltd
Donkin & Co. Ltd
Doxford Engines Ltd
Evershed & Vignoles Ltd

Fläkt Ltd (*SF Review*)
Foster Wheeler Power Products
  Ltd
Frydenbö Mek. Verksted
GEC Turbine Generators Ltd,
  Industrial & Marine Steam
  Turbine Division
Glacier Metal Co. Ltd
Grandi Motori Trieste
Graviner Ltd
M. W. Grazebook Ltd
Hall-Thermotank International
  Ltd
Hall-Thermotank Products Ltd
Hamworthy Combustion Systems
  Ltd
Hamworthy Engineering Ltd
Howaldtswerke-Deutsche Werft
John Hastie of Greenock Ltd
Richard Klinger Ltd
Maag Gearwheel Co. Ltd
McGregor Centrex Ltd
H. Maihak AG
Mather & Platt (Marine Dept.) Ltd

Michell Bearings Ltd
Mitsubishi Heavy Industries Ltd
The Motor Ship
NEI-APE Ltd
Nife Jungner AB, A/S
Norsk Elektrisk & Brown Boveri
Nu-Swift International Ltd
Peabody Holmes Ltd
Pyropress Engineering Co. Ltd
Scanpump AB
SEMT Pielstick
Serck Heat Transfer
Shipbuilding and Marine Engineering International
Siebe Gorman & Co. Ltd
Spirax Sarco Ltd
Stone Manganese Marine Ltd
Sulzer Brothers Ltd

Taylor Instrument Ltd
Thom, Lamont & Co. Ltd
Thompson Cochran Boilers Ltd
The Trent Valve Co. Ltd
Tungsten Batteries Ltd
Vokes Ltd
Vulkan Kupplungs-U.
    Getriebebau B. Hackforth
    GmbH & Co. KG
Walter Kidde & Co. Ltd
Weir Pumps Ltd
The Welin Davit & Engineering
    Co. Ltd
Weser AG
Wilson Elsan Marine International
    Ltd
Worthington-Simpson Ltd
Young and Cunningham Ltd

Contents

Chapter 1
Ships and machinery

As an introduction to marine engineering, we might reasonably begin by taking an overall look at the ship. The various duties of a marine engineer all relate to the operation of the ship in a safe, reliable, efficient and economic manner. The main propulsion machinery installed will influence the machinery layout and determine the equipment and auxiliaries installed. This will further determine the operational and maintenance requirements for the ship and thus the knowledge required and the duties to be performed by the marine engineer.

Ships

Ships are large, complex vehicles which must be self-sustaining in their environment for long periods with a high degree of reliability. A ship is the product of two main areas of skill, those of the naval architect and the marine engineer. The naval architect is concerned with the hull, its construction, form, habitability and ability to endure its environment. The marine engineer is responsible for the various systems which propel and operate the ship. More specifically, this means the machinery required for propulsion, steering, anchoring and ship securing, cargo handling, air conditioning, power generation and its distribution. Some overlap in responsibilities occurs between naval architects and marine engineers in areas such as propeller design, the reduction of noise and vibration in the ship's structure, and engineering services provided to considerable areas of the ship.

A ship might reasonably be divided into three distinct areas: the cargo-carrying holds or tanks, the accommodation and the machinery space. Depending upon the type each ship will assume varying proportions and functions. An oil tanker, for instance, will have the cargo-carrying region divided into tanks by two longitudinal bulkheads and several transverse bulkheads. There will be considerable quantities of cargo piping both above and below decks. The general cargo ship will

1

have various cargo holds which are usually the full width of the vessel and formed by transverse bulkheads along the ship's length. Cargo-handling equipment will be arranged on deck and there will be large hatch openings closed with steel hatch covers. The accommodation areas in each of these ship types will be sufficient to meet the requirements for the ship's crew, provide a navigating bridge area and a communications centre. The machinery space size will be decided by the particular machinery installed and the auxiliary equipment necessary. A passenger ship, however, would have a large accommodation area, since this might be considered the 'cargo space'. Machinery space requirements will probably be larger because of air conditioning equipment, stabilisers and other passenger related equipment.

Machinery

Arrangement

Three principal types of machinery installation are to be found at sea today. Their individual merits change with technological advances and improvements and economic factors such as the change in oil prices. It is intended therefore only to describe the layouts from an engineering point of view. The three layouts involve the use of direct-coupled slow-speed diesel engines, medium-speed diesels with a gearbox, and the steam turbine with a gearbox drive to the propeller.

A propeller, in order to operate efficiently, must rotate at a relatively low speed. Thus, regardless of the rotational speed of the prime mover, the propeller shaft must rotate at about 80 to 100 rev/min. The slow-speed diesel engine rotates at this low speed and the crankshaft is thus directly coupled to the propeller shafting. The medium-speed diesel engine operates in the range 250–750 rev/min and cannot therefore be directly coupled to the propeller shaft. A gearbox is used to provide a low-speed drive for the propeller shaft. The steam turbine rotates at a very high speed, in the order of 6000 rev/min. Again, a gearbox must be used to provide a low-speed drive for the propeller shaft.

Slow-speed diesel

A cutaway drawing of a complete ship is shown in Figure 1.1. Here, in addition to the machinery space, can be seen the structure of the hull, the cargo tank areas together with the cargo piping and the deck machinery. The compact, complicated nature of the machinery installation can clearly be seen, with the two major items being the main engine and the cargo heating boiler.

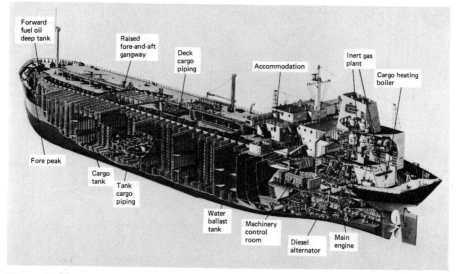

Figure 1.1 Cutaway drawing of a ship

The more usual plan and elevation drawings of a typical slow-speed diesel installation are shown in Figure 1.2.

A six-cylinder direct-drive diesel engine is shown in this machinery arrangement. The only auxiliaries visible are a diesel generator on the upper flat and an air compressor below. Other auxiliaries within the machinery space would include additional generators, an oily-water separator, an evaporator, numerous pumps and heat exchangers. An auxiliary boiler and an exhaust gas heat exchanger would be located in the uptake region leading to the funnel. Various workshops and stores and the machinery control room will also be found on the upper flats.

Geared medium-speed diesel

Four medium-speed (500 rev/min) diesels are used in the machinery layout of the rail ferry shown in Figure 1.3. The gear units provide a twin-screw drive at 170 rev/min to controllable-pitch propellers. The gear units also power take-offs for shaft-driven generators which provide all power requirements while at sea.

The various pumps and other auxiliaries are arranged at floor plate level in this minimum-height machinery space. The exhaust gas boilers and uptakes are located port and starboard against the side shell plating.

A separate generator room houses three diesel generator units, a waste combustion plant and other auxiliaries. The machinery control room is at the forward end of this room.

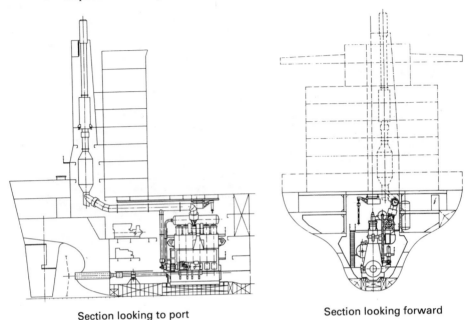

Section looking to port   Section looking forward

Figure 1.2 Slow-speed diesel machinery arrangement

Steam turbine

Twin cross-compounded steam turbines are used in the machinery layout of the container ship, shown in Figure 1.4. Only part plans and sections are given since there is a considerable degree of symmetry in the layout. Each turbine set drives, through a double reduction gearbox with separate thrust block, its own fixed-pitch propeller. The condensers are located beneath each low-pressure turbine and are arranged for scoop circulation at full power operation and axial pump circulation when manœuvring.

At the floorplate level around the main machinery are located various main engine and ship's services pumps, an auxiliary oil-fired boiler and a sewage plant. Three diesel alternators are located aft behind an acoustic screen.

The 8.5 m flat houses a turbo-alternator each side and also the forced-draught fans for the main boilers. The main boiler feed pumps and other feed system equipment are also located around this flat. The two main boilers occupy the after end of this flat and are arranged for roof firing. Two distillation plants are located forward and the domestic water supply units are located aft.

The control room is located forward of the 12.3 m flat and contains the main and auxiliary machinery consoles. The main switchboard and

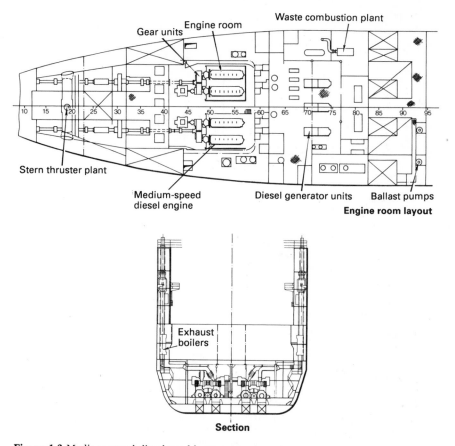

Figure 1.3 Medium-speed diesel machinery arrangement

group starter boards are located forward of the console, which faces into the machinery space.

On the 16.2 m flat is the combustion control equipment for each boiler with a local display panel, although control is from the main control room. The boiler fuel heating and pumping module is also located here. The de-aerator is located high up in the casing and silencers for the diesel alternators are in the funnel casing.

Operation and maintenance

The responsibilities of the marine engineer are rarely confined to the machinery space. Different companies have different practices, but usually all shipboard machinery, with the exception of radio equipment, is maintained by the marine engineer. Electrical engineers may be

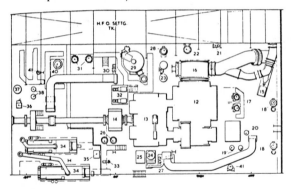

(a) *Part plan at floorplate level*

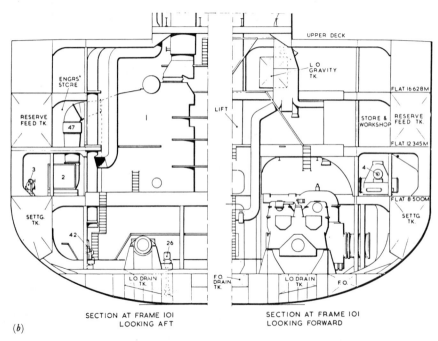

SECTION AT FRAME 101
LOOKING AFT

SECTION AT FRAME 101
LOOKING FORWARD

(b)

Figure 1.4 Steam turbine machinery arrangement

1 Main boiler	16 Main condenser	29 Auxiliary boiler
2 FD fan	17 Main extraction pump	30 Auxiliary boiler feed heater
3 Main feed pump	18 Bilge/ballast pump	31 HFO transfer pump module
4 Turbo-alternator	19 Drains tank extraction	32 HFO service pumps
7 SW-cooled evaporator	pumps	33 Diesel oil transfer pump
10 Hot water calorifier	21 Turbo alternator pump	34 Diesel alternator
11 FW pressure tank	22 LO cooler	35 Diesel alternator controls
12 Main turbines	24 LO bypass filter and	40 Condensate de-oiler
13 Main gearbox	pumps	41 Refrigerant circulation pump
14 Thrust block	26 LO pumps	42 Oily bilge pump
15 Main SW circ pump	28 Fire pump	43 Steam/air heater

carried on very large ships, but if not, the electrical equipment is also maintained by the engineer.

A broad-based theoretical and practical training is therefore necessary for a marine engineer. He must be a mechanical, electrical, air conditioning, ventilation and refrigeration engineer, as the need arises. Unlike his shore-based opposite number in these occupations, he must also deal with the specialised requirements of a floating platform in a most corrosive environment. Furthermore he must be self sufficient and capable of getting the job done with the facilities at his disposal.

The modern ship is a complex collection of self-sustaining machinery providing the facilities to support a small community for a considerable period of time. To simplify the understanding of all this equipment is the purpose of this book. This equipment is dealt with either as a complete system comprising small items or individual larger items. In the latter case, especially, the choices are often considerable. A knowledge of machinery and equipment operation provides the basis for effective maintenance, and the two are considered in turn in the following chapters.

Chapter 2
Diesel engines

The diesel engine is a type of internal combustion engine which ignites the fuel by injecting it into hot, high-pressure air in a combustion chamber. In common with all internal combustion engines the diesel engine operates with a fixed sequence of events, which may be achieved either in four strokes or two, a stroke being the travel of the piston between its extreme points. Each stroke is accomplished in half a revolution of the crankshaft.

Four-stroke cycle

The four-stroke cycle is completed in four strokes of the piston, or two revolutions of the crankshaft. In order to operate this cycle the engine requires a mechanism to open and close the inlet and exhaust valves.

Consider the piston at the top of its stroke, a position known as top dead centre (TDC). The inlet valve opens and fresh air is drawn in as the piston moves down (Figure 2.1(a)). At the bottom of the stroke, i.e. bottom dead centre (BDC), the inlet valve closes and the air in the cylinder is compressed (and consequently raised in temperature) as the piston rises (Figure 2.1(b)). Fuel is injected as the piston reaches top dead centre and combustion takes place, producing very high pressure in the gases (Figure 2.1(c)). The piston is now forced down by these gases and at bottom dead centre the exhaust valve opens. The final stroke is the exhausting of the burnt gases as the piston rises to top dead centre to complete the cycle (Figure 2.1(d)). The four distinct strokes are known as 'inlet' (or suction), 'compression', 'power' (or working stroke) and 'exhaust'.

These events are shown diagrammatically on a timing diagram (Figure 2.2). The angle of the crank at which each operation takes place is shown as well as the period of the operation in degrees. This diagram is more correctly representative of the actual cycle than the simplified explanation given in describing the four-stroke cycle. For different engine designs the different angles will vary, but the diagram is typical.

8

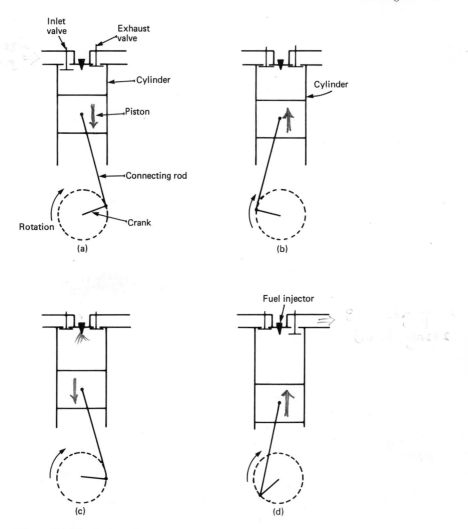

Figure 2.1 The four-stroke cycle. (a) suction stroke and (b) compression stroke. (c) power stroke and (d) exhaust stroke

Two-stroke cycle

The two-stroke cycle is completed in two strokes of the piston or one revolution of the crankshaft. In order to operate this cycle where each event is accomplished in a very short time, the engine requires a number of special arrangements. First, the fresh air must be forced in under pressure. The incoming air is used to clean out or scavenge the exhaust

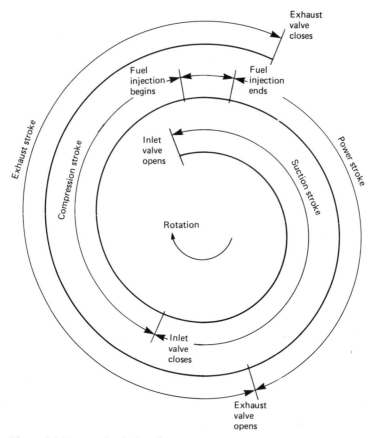

Figure 2.2 Four-stroke timing diagram

gases and then to fill or charge the space with fresh air. Instead of valves holes, known as 'ports', are used which are opened and closed by the sides of the piston as it moves.

Consider the piston at the top of its stroke where fuel injection and combustion have just taken place (Figure 2.3(a)). The piston is forced down on its working stroke until it uncovers the exhaust port (Figure 2.3(b)). The burnt gases then begin to exhaust and the piston continues down until it opens the inlet or scavenge port (Figure 2.3(c)). Pressurised air then enters and drives out the remaining exhaust gas. The piston, on its return stroke, closes the inlet and exhaust ports. The air is then compressed as the piston moves to the top of its stroke to complete the cycle (Figure 2.3(d)). A timing diagram for a two-stroke engine is shown in Figure 2.4.

The opposed piston cycle of operations is a special case of the two-stroke cycle. Beginning at the moment of fuel injection, both pistons

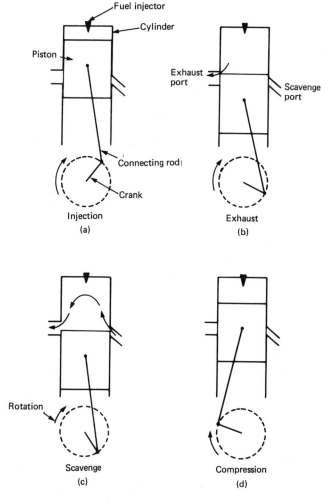

Figure 2.3 Two-stroke cycle

are forced apart—one up, one down—by the expanding gases (Figure 2.5(a)). The upper piston opens the exhaust ports as it reaches the end of its travel (Figure 2.5(b)). The lower piston, a moment or two later, opens the scavenge ports to charge the cylinder with fresh air and remove the final traces of exhaust gas (Figure 2.5(c)). Once the pistons reach their extreme points they both begin to move inward. This closes off the scavenge and exhaust ports for the compression stroke to take place prior to fuel injection and combustion (Figure 2.5(d)). This cycle is used in the Doxford engine, which is no longer manufactured although many are still in operation.

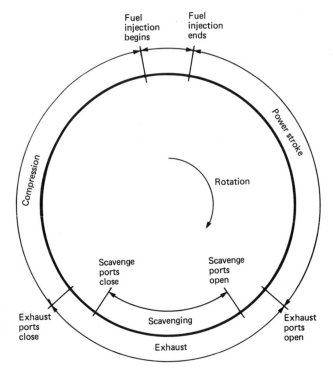

Figure 2.4 Two-stroke timing diagram

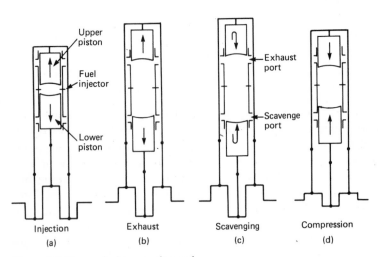

Figure 2.5 Opposed piston engine cycle

The four-stroke engine

A cross-section of a four-stroke cycle engine is shown in Figure 2.6. The engine is made up of a piston which moves up and down in a cylinder which is covered at the top by a cylinder head. The fuel injector, through which fuel enters the cylinder, is located in the cylinder head. The inlet and exhaust valves are also housed in the cylinder head and held shut by springs. The piston is joined to the connecting rod by a gudgeon pin. The bottom end or big end of the connecting rod is joined to the crankpin which forms part of the crankshaft. With this assembly the

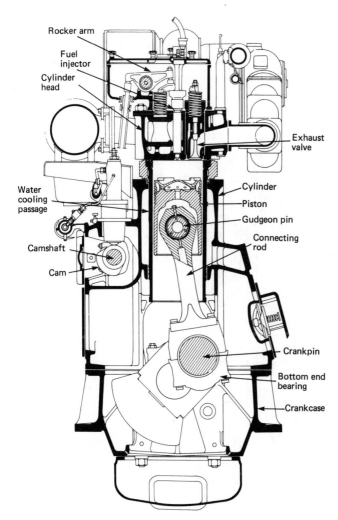

Figure 2.6 Cross-section of a four-stroke diesel engine

linear up-and-down movement of the piston is converted into rotary movement of the crankshaft. The crankshaft is arranged to drive through gears the camshaft, which either directly or through pushrods operates rocker arms which open the inlet and exhaust valves. The camshaft is 'timed' to open the valves at the correct point in the cycle. The crankshaft is surrounded by the crankcase and the engine framework which supports the cylinders and houses the crankshaft bearings. The cylinder and cylinder head are arranged with water-cooling passages around them.

The two-stroke engine

A cross-section of a two-stroke cycle engine is shown in Figure 2.7. The piston is solidly connected to a piston rod which is attached to a crosshead bearing at the other end. The top end of the connecting rod is

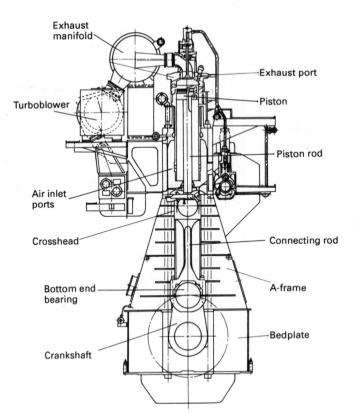

Figure 2.7 Cross-section of a two-stroke diesel engine

also joined to the crosshead bearing. Ports are arranged in the cylinder liner for air inlet and a valve in the cylinder head enables the release of exhaust gases. The incoming air is pressurised by a turbo-blower which is driven by the outgoing exhaust gases. The crankshaft is supported within the engine bedplate by the main bearings. A-frames are mounted on the bedplate and house guides in which the crosshead travels up and down. The entablature is mounted above the frames and is made up of the cylinders, cylinder heads and the scavenge trunking.

Comparison of two-stroke and four-stroke cycles

The main difference between the two cycles is the power developed. The two-stroke cycle engine, with one working or power stroke every revolution, will, theoretically, develop twice the power of a four-stroke engine of the same swept volume. Inefficient scavenging however and other losses, reduce the power advantage to about 1.8. For a particular engine power the two-stroke engine will be considerably lighter—an important consideration for ships. Nor does the two-stroke engine require the complicated valve operating mechanism of the four-stroke. The four-stroke engine however can operate efficiently at high speeds which offsets its power disadvantage; it also consumes less lubricating oil.

Each type of engine has its applications which on board ship have resulted in the slow speed (i.e. 80–100 rev/min) main propulsion diesel operating on the two-stroke cycle. At this low speed the engine requires no reduction gearbox between it and the propeller. The four-stroke engine (usually rotating at medium speed, between 250 and 750 rev/ min) is used for auxiliaries such as alternators and sometimes for main propulsion with a gearbox to provide a propeller speed of between 80 and 100 rev/min.

Power measurement

There are two possible measurements of engine power: the *indicated power* and the *shaft power*. The indicated power is the power developed within the engine cylinder and can be measured by an engine indicator. The shaft power is the power available at the output shaft of the engine and can be measured using a torsionmeter or with a brake.

The engine indicator

An engine indicator is shown in Figure 2.8. It is made up of a small piston of known size which operates in a cylinder against a specially

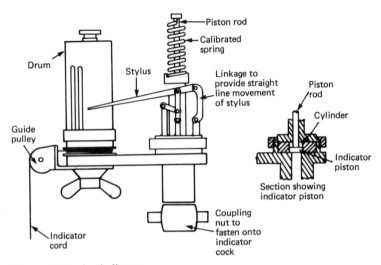

Figure 2.8 Engine indicator

calibrated spring. A magnifying linkage transfers the piston movement to a drum on which is mounted a piece of paper or card. The drum oscillates (moves backwards and forwards) under the pull of the cord. The cord is moved by a reciprocating (up and down) mechanism which is proportional to the engine piston movement in the cylinder. The stylus draws out an indicator diagram which represents the gas pressure on the engine piston at different points of the stroke, and the area of the indicator diagram produced represents the power developed in the particular cylinder. The cylinder power can be measured if the scaling factors, spring calibration and some basic engine details are known. The procedure is described in the Appendix. The cylinder power values are compared, and for balanced loading should all be the same. Adjustments may then be made to the fuel supply in order to balance the cylinder loads.

Torsionmeter

If the torque transmitted by a shaft is known, together with the angular velocity, then the power can be measured, i.e.

shaft power = torque × angular velocity

The torque on a shaft can be found by measuring the shear stress or angle of twist with a torsionmeter. A number of different types of torsionmeter are described in Chapter 15.

The gas exchange process

A basic part of the cycle of an internal combustion engine is the supply of fresh air and removal of exhaust gases. This is the gas exchange process. *Scavenging* is the removal of exhaust gases by blowing in fresh air. *Charging* is the filling of the engine cylinder with a supply or charge of fresh air ready for compression. With *supercharging* a large mass of air is supplied to the cylinder by blowing it in under pressure. Older engines were 'naturally aspirated'—taking fresh air only at atmospheric pressure. Modern engines make use of exhaust gas driven turbochargers to supply pressurised fresh air for scavenging and supercharging. Both four-stroke and two-stroke cycle engines may be pressure charged.

On two-stroke diesels an electrically driven auxiliary blower is usually provided because the exhaust gas driven turboblower cannot provide enough air at low engine speeds, and the pressurised air is usually cooled to increase the charge air density. An exhaust gas driven turbocharging arrangement for a slow-speed two-stroke cycle diesel is shown in Figure 2.9(a).

A turboblower or turbocharger is an air compressor driven by exhaust gas (Figure 2.9(b)). The single shaft has an exhaust gas turbine on one end and the air compressor on the other. Suitable casing design and shaft seals ensure that the two gases do not mix. Air is drawn from the machinery space through a filter and then compressed before passing to the scavenge space. The exhaust gas may enter the turbine directly from the engine or from a constant-pressure chamber. Each of the shaft bearings has its own independent lubrication system, and the exhaust gas end of the casing is usually water-cooled.

Scavenging

Efficient scavenging is essential to ensure a sufficient supply of fresh air for combustion. In the four-stroke cycle engine there is an adequate overlap between the air inlet valve opening and the exhaust valve closing. With two-stroke cycle engines this overlap is limited and some slight mixing of exhaust gases and incoming air does occur.

A number of different scavenging methods are in use in slow-speed two-stroke engines. In each the fresh air enters as the inlet port is opened by the downward movement of the piston and continues until the port is closed by the upward moving piston. The flow path of the scavenge air is decided by the engine port shape and design and the exhaust arrangements. Three basic systems are in use: the cross flow, the loop and the uniflow. All modern slow-speed diesel engines now use the uniflow scavenging system with a cylinder-head exhaust valve.

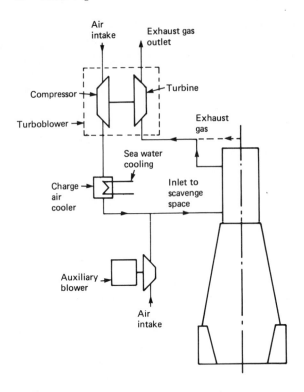

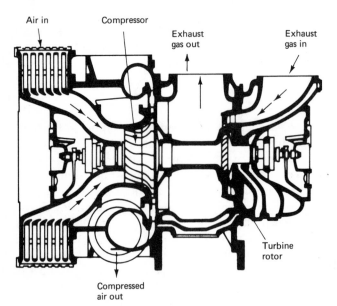

Figure 2.9 (a) Exhaust gas turbocharging arrangement. (b) A turbocharger

In cross scavenging the incoming air is directed upwards, pushing the exhaust gases before it. The exhaust gases then travel down and out of the exhaust ports. Figure 2.10(a) illustrates the process.

In loop scavenging the incoming air passes over the piston crown then rises towards the cylinder head. The exhaust gases are forced before the air passing down and out of exhaust ports located just above the inlet ports. The process is shown in Figure 2.10(b).

With uniflow scavenging the incoming air enters at the lower end of the cylinder and leaves at the top. The outlet at the top of the cylinder may be ports or a large valve. The process is shown in Figure 2.10(c).

Each of the systems has various advantages and disadvantages. Cross scavenging requires the fitting of a piston skirt to prevent air or exhaust gas escape when the piston is at the top of the stroke. Loop scavenge

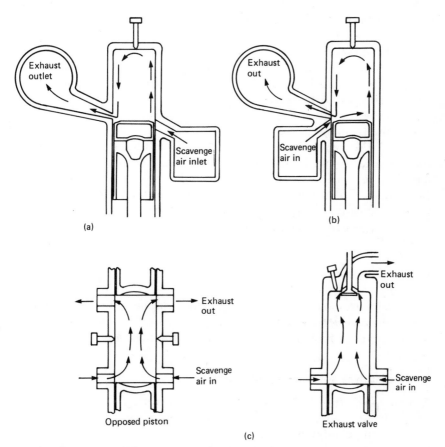

Figure 2.10 Scavenging methods. (a) Cross-flow scavenging, (b) loop scavenging, (c) uniflow scavenging

arrangements have low temperature air and high temperature exhaust gas passing through adjacent ports, causing temperature differential problems for the liner material. Uniflow is the most efficient scavenging system but requires either an opposed piston arrangement or an exhaust valve in the cylinder head. All three systems have the ports angled to swirl the incoming air and direct it in the appropriate path.

Scavenge fires

Cylinder oil can collect in the scavenge space of an engine. Unburned fuel and carbon may also be blown into the scavenge space as a result of defective piston rings, faulty timing, a defective injector, etc. A build-up of this flammable mixture presents a danger as a blow past of hot gases from the cylinder may ignite the mixture, and cause a scavenge fire.

A loss of engine power will result, with high exhaust temperatures at the affected cylinders. The affected turbo-chargers may surge and sparks will be seen at the scavenge drains. Once a fire is detected the engine should be slowed down, fuel shut off from the affected cylinders and cylinder lubrication increased. All the scavenge drains should be closed. A small fire will quickly burn out, but where the fire persists the engine must be stopped. A fire extinguishing medium should then be injected through the fittings provided in the scavenge trunking. On no account should the trunking be opened up.

To avoid scavenge fires occurring the engine timing and equipment maintenance should be correctly carried out. The scavenge trunking should be regularly inspected and cleaned if necessary. Where carbon or oil build up is found in the scavenge, its source should be detected and the fault remedied. Scavenge drains should be regularly blown and any oil discharges investigated at the first opportunity.

Fuel oil system

The fuel oil system for a diesel engine can be considered in two parts—the *fuel supply* and the *fuel injection* systems. Fuel supply deals with the provision of fuel oil suitable for use by the injection system.

Fuel oil supply for a two-stroke diesel

A slow-speed two-stroke diesel is usually arranged to operate continuously on heavy fuel and have available a diesel oil supply for manœuvring conditions.

In the system shown in Figure 2.11, the oil is stored in tanks in the double bottom from which it is pumped to a settling tank and heated.

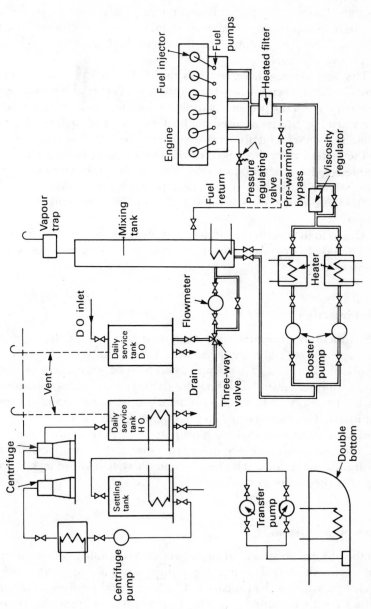

Figure 2.11 Fuel oil supply system

After passing through centrifuges the cleaned, heated oil is pumped to a daily service tank. From the daily service tank the oil flows through a three-way valve to a mixing tank. A flow meter is fitted into the system to indicate fuel consumption. Booster pumps are used to pump the oil through heaters and a viscosity regulator to the engine-driven fuel pumps. The fuel pumps will discharge high-pressure fuel to their respective injectors.

The viscosity regulator controls the fuel oil temperature in order to provide the correct viscosity for combustion. A pressure regulating valve ensures a constant-pressure supply to the engine-driven pumps, and a pre-warming bypass is used to heat up the fuel before starting the engine. A diesel oil daily service tank may be installed and is connected to the system via a three-way valve. The engine can be started up and manœuvred on diesel oil or even a blend of diesel and heavy fuel oil. The mixing tank is used to collect recirculated oil and also acts as a buffer or reserve tank as it will supply fuel when the daily service tank is empty.

The system includes various safety devices such as low-level alarms and remotely operated tank outlet valves which can be closed in the event of a fire.

Fuel injection

The function of the fuel injection system is to provide the right amount of fuel at the right moment and in a suitable condition for the combustion process. There must therefore be some form of measured fuel supply, a means of timing the delivery and the atomisation of the fuel. The injection of the fuel is achieved by the location of cams on a camshaft. This camshaft rotates at engine speed for a two-stroke engine and at half engine speed for a four-stroke. There are two basic systems in use, each of which employs a combination of mechanical and hydraulic operations. The most common system is the jerk pump: the other is the common rail.

Jerk pump system

In the jerk pump system of fuel injection a separate injector pump exists for each cylinder. The injector pump is usually operated once every cycle by a cam on the camshaft. The barrel and plunger of the injector pump are dimensioned to suit the engine fuel requirements. Ports in the barrel and slots in the plunger or adjustable spill valves serve to regulate the fuel delivery (a more detailed explanation follows). Each injector pump supplies the injector or injectors for one cylinder. The needle valve in the injector will lift at a pre-set pressure which ensures that the fuel will atomise once it enters the cylinder.

There are two particular types of fuel pump in use, the valve-controlled discharge type and the helix or helical edge pump. Valve-controlled pumps are used on slow-speed two-stroke engines and the helix type for all medium- and high-speed four-stroke engines.

Helix-type injector pump

The injector pump is operated by a cam which drives the plunger up and down. The timing of the injection can be altered by raising or lowering the pump plunger in relation to the cam. The pump has a constant stroke and the amount of fuel delivered is regulated by rotating the pump plunger which has a specially arranged helical groove cut into it.

The fuel is supplied to the pump through ports or openings at B (Figure 2.12). As the plunger moves down, fuel enters the cylinder. As the plunger moves up, the ports at B are closed and the fuel is pressurised and delivered to the injector nozzle at very high pressure. When the edge of the helix at C uncovers the spill port D pressure is lost and fuel delivery to the injector stops. A non-return valve on the delivery side of the pump closes to stop fuel oil returning from the injector. Fuel will again be drawn in on the plunger downstroke and the process will be repeated.

The plunger may be rotated in the cylinder by a rack and pinion arrangement on a sleeve which is keyed to the plunger. This will move the edge C up or down to reduce or increase the amount of fuel pumped into the cylinder. The rack is connected to the throttle control or governor of the engine.

This type of pump, with minor variations, is used on many four-stroke diesel engines.

Valve-controlled pump

In the variable injection timing (VIT) pump used in MAN B&W engines the governor output shaft is the controlling parameter. Two linkages are actuated by the regulating shaft of the governor.

The upper control linkage changes the injection timing by raising or lowering the plunger in relation to the cam. The lower linkage rotates the pump plunger and thus the helix in order to vary the pump output (Figure 2.13).

In the Sulzer variable injection timing system the governor output is connected to a suction valve and a spill valve. The closing of the pump suction valve determines the beginning of injection. Operation of the

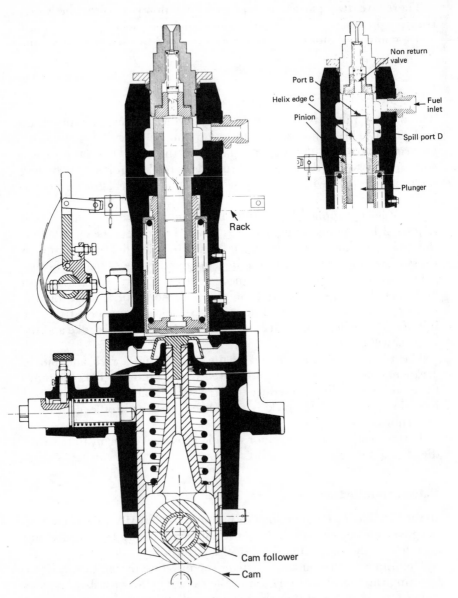

Figure 2.12 Injector pump with detail view showing ports and plunger

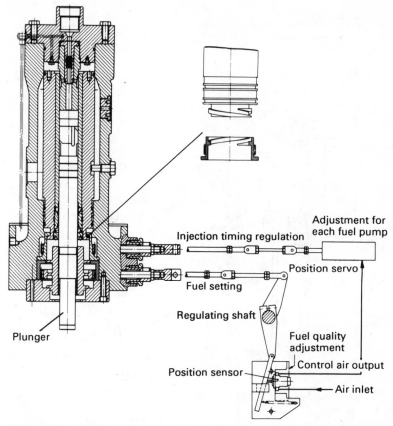

Figure 2.13 Variable injection timing (VIT) pump

spill valve will control the end of injection by releasing fuel pressure. No helix is therefore present on the pump plunger.

Common rail system

The common rail system has one high-pressure multiple plunger fuel pump (Figure 2.14). The fuel is discharged into a manifold or rail which is maintained at high pressure. From this common rail fuel is supplied to all the injectors in the various cylinders. Between the rail and the injector or injectors for a particular cylinder is a timing valve which determines the timing and extent of fuel delivery. Spill valves are connected to the manifold or rail to release excess pressure and accumulator bottles which dampen out pump pressure pulses. The injectors in a common rail system are often referred to as fuel valves.

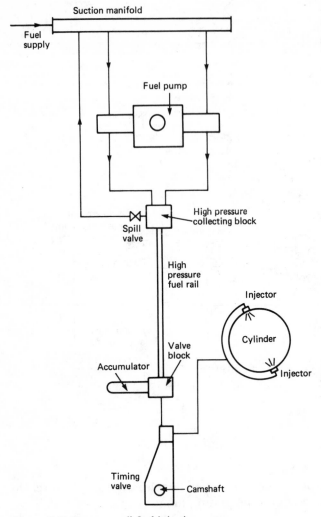

Figure 2.14 Common rail fuel injection system

Timing valve

The timing valve in the common rail system is operated by a cam and lever (Figure 2.15). When the timing valve is lifted by the cam and lever the high-pressure fuel flows to the injector. The timing valve operating lever is fixed to a sliding rod which is positioned according to the manœuvring lever setting to provide the correct fuel quantity to the cylinder.

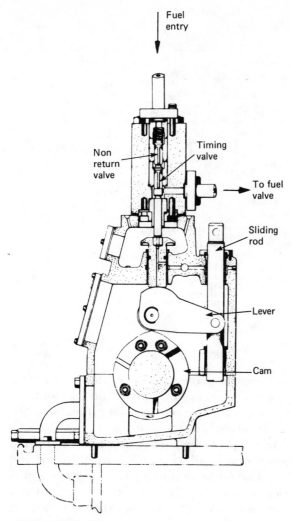

Figure 2.15 Timing valve

The fuel injector

A typical fuel injector is shown in Figure 2.16. It can be seen to be two
basic parts, the nozzle and the nozzle holder or body. The high-pressure
fuel enters and travels down a passage in the body and then into a
passage in the nozzle, ending finally in a chamber surrounding the
needle valve. The needle valve is held closed on a mitred seat by an
intermediate spindle and a spring in the injector body. The spring

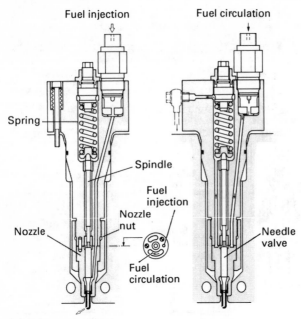

Figure 2.16 Fuel injector

pressure, and hence the injector opening pressure, can be set by a compression nut which acts on the spring. The nozzle and injector body are manufactured as a matching pair and are accurately ground to give a good oil seal. The two are joined by a nozzle nut.

The needle valve will open when the fuel pressure acting on the needle valve tapered face exerts a sufficient force to overcome the spring compression. The fuel then flows into a lower chamber and is forced out through a series of tiny holes. The small holes are sized and arranged to atomise, or break into tiny drops, all of the fuel oil, which will then readily burn. Once the injector pump or timing valve cuts off the high pressure fuel supply the needle valve will shut quickly under the spring compression force.

All slow-speed two-stroke engines and many medium-speed four-stroke engines are now operated almost continuously on heavy fuel. A fuel circulating system is therefore necessary and this is usually arranged within the fuel injector. During injection the high-pressure fuel will open the circulation valve for injection to take place. When the engine is stopped the fuel booster pump supplies fuel which the circulation valve directs around the injector body.

Older engine designs may have fuel injectors which are circulated with cooling water.

Lubrication

The lubrication system of an engine provides a supply of lubricating oil to the various moving parts in the engine. Its main function is to enable the formation of a film of oil between the moving parts, which reduces friction and wear. The lubricating oil is also used as a cleaner and in some engines as a coolant.

Lubricating oil system

Lubricating oil for an engine is stored in the bottom of the crankcase, known as the sump, or in a drain tank located beneath the engine (Figure 2.17). The oil is drawn from this tank through a strainer, one of a pair of pumps, into one of a pair of fine filters. It is then passed through a cooler before entering the engine and being distributed to the various branch pipes. The branch pipe for a particular cylinder may feed the main bearing, for instance. Some of this oil will pass along a drilled passage in the crankshaft to the bottom end bearing and then up a drilled passage in the connecting rod to the gudgeon pin or crosshead bearing. An alarm at the end of the distribution pipe ensures that adequate pressure is maintained by the pump. Pumps and fine filters are

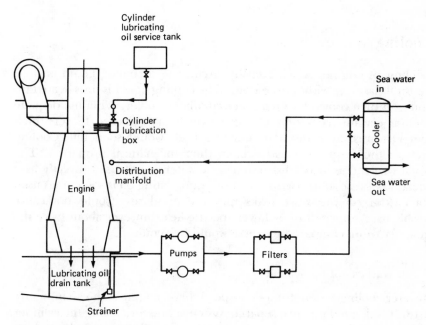

Figure 2.17 Lubricating oil system

arranged in duplicate with one as standby. The fine filters will be arranged so that one can be cleaned while the other is operating. After use in the engine the lubricating oil drains back to the sump or drain tank for re-use. A level gauge gives a local read-out of the drain tank contents. A centrifuge is arranged for cleaning the lubricating oil in the system and clean oil can be provided from a storage tank.

The oil cooler is circulated by sea water, which is at a lower pressure than the oil. As a result any leak in the cooler will mean a loss of oil and not contamination of the oil by sea water.

Where the engine has oil-cooled pistons they will be supplied from the lubricating oil system, possibly at a higher pressure produced by booster pumps, e.g. Sulzer RTA engine. An appropriate type of lubricating oil must be used for oil-lubricated pistons in order to avoid carbon deposits on the hotter parts of the system.

Cylinder lubrication

Large slow-speed diesel engines are provided with a separate lubrication system for the cylinder liners. Oil is injected between the liner and the piston by mechanical lubricators which supply their individual cylinder. A special type of oil is used which is not recovered. As well as lubricating, it assists in forming a gas seal and contains additives which clean the cylinder liner.

Cooling

Cooling of engines is achieved by circulating a cooling liquid around internal passages within the engine. The cooling liquid is thus heated up and is in turn cooled by a sea water circulated cooler. Without adequate cooling certain parts of the engine which are exposed to very high temperatures, as a result of burning fuel, would soon fail. Cooling enables the engine metals to retain their mechanical properties. The usual coolant used is fresh water: sea water is not used directly as a coolant because of its corrosive action. Lubricating oil is sometimes used for piston cooling since leaks into the crankcase would not cause problems. As a result of its lower specific heat however about twice the quantity of oil compared to water would be required.

Fresh water cooling system

A water cooling system for a slow-speed diesel engine is shown in Figure 2.18. It is divided into two separate systems: one for cooling the cylinder jackets, cylinder heads and turbo-blowers; the other for piston cooling.

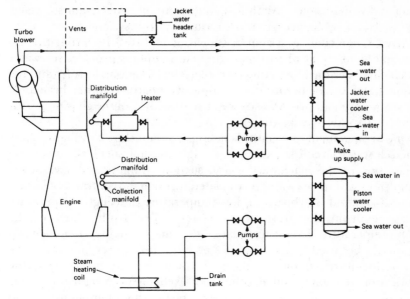

Figure 2.18 Fresh water cooling system

The cylinder jacket cooling water after leaving the engine passes to a sea-water-circulated cooler and then into the jacket-water circulating pumps. It is then pumped around the cylinder jackets, cylinder heads and turbo-blowers. A header tank allows for expansion and water make-up in the system. Vents are led from the engine to the header tank for the release of air from the cooling water. A heater in the circuit facilitates warming of the engine prior to starting by circulating hot water.

The piston cooling system employs similar components, except that a drain tank is used instead of a header tank and the vents are then led to high points in the machinery space. A separate piston cooling system is used to limit any contamination from piston cooling glands to the piston cooling system only.

Sea water cooling system

The various cooling liquids which circulate the engine are themselves cooled by sea water. The usual arrangement uses individual coolers for lubricating oil, jacket water, and the piston cooling system, each cooler being circulated by sea water. Some modern ships use what is known as a 'central cooling system' with only one large sea-water-circulated cooler. This cools a supply of fresh water, which then circulates to the other individual coolers. With less equipment in contact with sea water the corrosion problems are much reduced in this system.

A sea water cooling system is shown in Figure 2.19. From the sea suction one of a pair of sea-water circulating pumps provides sea water which circulates the lubricating oil cooler, the jacket water cooler and the piston water cooler before discharging overboard. Another branch of the sea water main provides sea water to directly cool the charge air (for a direct-drive two-stroke diesel).

One arrangement of a central cooling system is shown in Figure 2.20. The sea water circuit is made up of high and low suctions, usually on either side of the machinery space, suction strainers and several sea water pumps. The sea water is circulated through the central coolers and then discharged overboard. A low-temperature and high-temperature circuit exist in the fresh water system. The fresh water in the high-temperature circuit circulates the main engine and may, if required, be used as a heating medium for an evaporator. The low-temperature circuit circulates the main engine air coolers, the lubricating oil coolers and all other heat exchangers. A regulating valve controls the mixing of water between the high-temperature and low-temperature circuits. A temperature sensor provides a signal to the

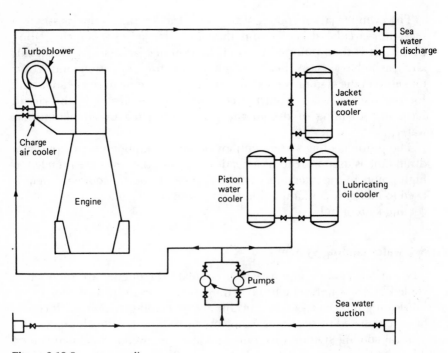

Figure 2.19 Sea water cooling system

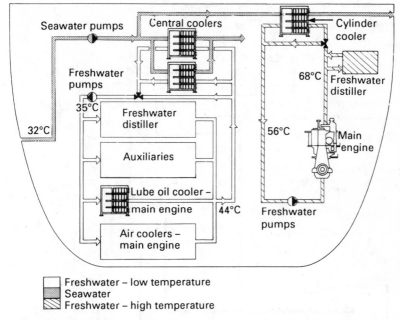

Freshwater – low temperature
Seawater
Freshwater – high temperature

Figure 2.20 Central cooling system

control unit which operates the regulating valve to maintain the desired temperature setting. A temperature sensor is also used in a similar control circuit to operate the regulating valve which controls the bypassing of the central coolers.

It is also possible, with appropriate control equipment, to vary the quantity of sea water circulated by the pumps to almost precisely meet the cooler requirements.

Starting air system

Diesel engines are started by supplying compressed air into the cylinders in the appropriate sequence for the required direction. A supply of compressed air is stored in air reservoirs or 'bottles' ready for immediate use. Up to 12 starts are possible with the stored quantity of compressed air. The starting air system usually has interlocks to prevent starting if everything is not in order.

A starting air system is shown in Figure 2.21. Compressed air is supplied by air compressors to the air receivers. The compressed air is then supplied by a large bore pipe to a remote operating non-return or automatic valve and then to the cylinder air start valve. Opening of the

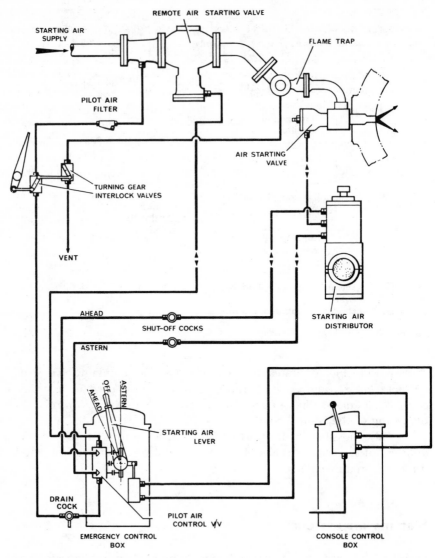

Figure 2.21 Starting air system

cylinder air start valve will admit compressed air into the cylinder. The opening of the cylinder valve and the remote operating valve is controlled by a pilot air system. The pilot air is drawn from the large pipe and passes to a pilot air control valve which is operated by the engine air start lever.

When the air start lever is operated, a supply of pilot air enables the remote valve to open. Pilot air for the appropriate direction of operation is also supplied to an air distributor. This device is usually driven by the engine camshaft and supplies pilot air to the control cylinders of the cylinder air start valves. The pilot air is then supplied in the appropriate sequence for the direction of operation required. The cylinder air start valves are held closed by springs when not in use and opened by the pilot air enabling the compressed air direct from the receivers to enter the engine cylinder. An interlock is shown in the remote operating valve line which stops the valve opening when the engine turning gear is engaged. The remote operating valve prevents the return of air which has been further compressed by the engine into the system.

Lubricating oil from the compressor will under normal operation pass along the air lines and deposit on them. In the event of a cylinder air starting valve leaking, hot gases would pass into the air pipes and ignite the lubricating oil. If starting air is supplied to the engine this would further feed the fire and could lead to an explosion in the pipelines. In order to prevent such an occurrence, cylinder starting valves should be properly maintained and the pipelines regularly drained. Also oil discharged from compressors should be kept to a minimum, by careful maintenance.

In an attempt to reduce the effects of an explosion, flame traps, relief valves and bursting caps or discs are fitted to the pipelines. In addition an isolating non-return valve (the automatic valve) is fitted to the system. The loss of cooling water from an air compressor could lead to an overheated air discharge and possibly an explosion in the pipelines leading to the air reservoir. A high-temperature alarm or a fusible plug which will melt is used to guard against this possibility.

Control and safety devices

Governors

The principal control device on any engine is the governor. It governs or controls the engine speed at some fixed value while power output changes to meet demand. This is achieved by the govenor automatically adjusting the engine fuel pump settings to meet the desired load at the set speed. Governors for diesel engines are usually made up of two systems: a speed sensing arrangement and a hydraulic unit which operates on the fuel pumps to change the engine power output.

Mechanical governor

A flyweight assembly is used to detect engine speed. Two flyweights are fitted to a plate or ballhead which rotates about a vertical axis driven by a gear wheel (Figure 2.22). The action of centrifugal force throws the

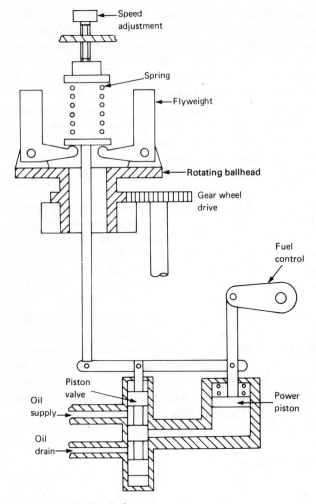

Figure 2.22 Mechanical governor

weights outwards; this lifts the vertical spindle and compresses the spring until an equilibrium situation is reached. The equilibrium position or set speed may be changed by the speed selector which alters the spring compression.

As the engine speed increases the weights move outwards and raise the spindle; a speed decrease will lower the spindle.

The hydraulic unit is connected to this vertical spindle and acts as a power source to move the engine fuel controls. A piston valve connected to the vertical spindle supplies or drains oil from the power piston which moves the fuel controls depending upon the flyweight movement. If the engine speed increases the vertical spindle rises, the piston valve rises and oil is drained from the power piston which results in a fuel control movement. This reduces fuel supply to the engine and slows it down. It is, in effect, a proportional controller (see Chapter 15).

The actual arrangement of mechanical engine governors will vary considerably but most will operate as described above.

Electric governor

The electric governor uses a combination of electrical and mechanical components in its operation. The speed sensing device is a small magnetic pick-up coil. The rectified, or d.c., voltage signal is used in conjunction with a desired or set speed signal to operate a hydraulic unit. This unit will then move the fuel controls in the appropriate direction to control the engine speed.

Cylinder relief valve

The cylinder relief valve is designed to relieve pressures in excess of 10% to 20% above normal. A spring holds the valve closed and its lifting pressure is set by an appropriate thickness of packing piece (Figure 2.23). Only a small amount of lift is permitted and the escaping gases are directed to a safe outlet. The valve and spindle are separate to enable the valve to correctly seat itself after opening.

The operation of this device indicates a fault in the engine which should be discovered and corrected. The valve itself should then be examined at the earliest opportunity.

Crankcase oil mist detector

The presence of an oil mist in the crankcase is the result of oil vaporisation caused by a hot spot. Explosive conditions can result if a build up of oil mist is allowed. The oil mist detector uses photoelectric cells to measure small increases in oil mist density. A motor driven fan continuously draws samples of crankcase oil mist through a measuring tube. An increased meter reading and alarm will result if any crankcase sample contains excessive mist when compared to either clean air or the other crankcase compartments. The rotary valve which draws the sample then stops to indicate the suspect crankcase. The *comparator* model tests one crankcase mist sample against all the others and once a cycle against clean air. The *level* model tests each crankcase in turn against a reference tube sealed with clean air. The comparator model is used for crosshead type engines and the level model for trunk piston engines.

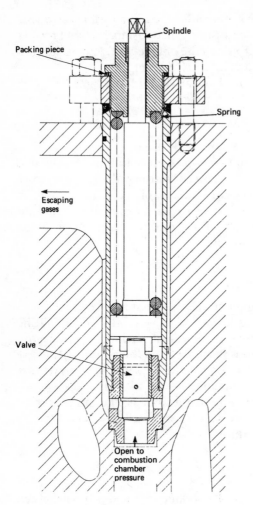

Spindle

Packing piece

Spring

Escaping
gases

Valve

Open to
combustion
chamber
pressure

Figure 2.23 Cylinder relief valve

Explosion relief valve

As a practical safeguard against explosions which occur in a crankcase, explosion relief valves or doors are fitted. These valves serve to relieve excessive crankcase pressures and stop flames being emitted from the crankcase. They must also be self closing to stop the return of atmospheric air to the crankcase.

Various designs and arrangements of these valves exist where, on large slow-speed diesels, two door type valves may be fitted to each crankcase or, on a medium-speed diesel, one valve may be used. One design of explosion relief valve is shown in Figure 2.24. A light spring

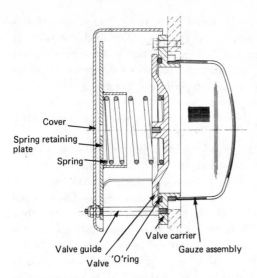

Cover

Spring retaining
plate

Spring

Valve guide

Valve

'O'ring

Valve carrier

Gauze assembly

Figure 2.24 Crankcase explosion relief valve

holds the valve closed against its seat and a seal ring completes the joint. A deflector is fitted on the outside of the engine to safeguard personnel from the outflowing gases, and inside the engine, over the valve opening, an oil wetted gauze acts as a flame trap to stop any flames leaving the crankcase. After operation the valve will close automatically under the action of the spring.

Turning gear

The turning gear or turning engine is a reversible electric motor which drives a worm gear which can be connected with the toothed flywheel to turn a large diesel. A slow-speed drive is thus provided to enable positioning of the engine parts for overhaul purposes. The turning gear is also used to turn the engine one or two revolutions prior to starting. This is a safety check to ensure that the engine is free to turn and that no water has collected in the cylinders. The indicator cocks must always be open when the turning gear is operated.

Medium- and slow-speed diesels

Medium-speed diesels, e.g. 250 to 750 rev/min, and slow-speed diesels, e.g. 100 to 120 rev/min, each have their various advantages and disadvantages for various duties on board ship.

The slow-speed two-stroke cycle diesel is used for main propulsion units since it can be directly coupled to the propeller and shafting. It provides high powers, can burn low-grade fuels and has a high thermal efficiency. The cylinders and crankcase are isolated, which reduces contamination and permits the use of specialised lubricating oils in each area. The use of the two-stroke cycle usually means there are no inlet and exhaust valves. This reduces maintenance and simplifies engine construction.

Medium-speed four-stroke engines provide a better power-to-weight ratio and power-to-size ratio and there is also a lower initial cost for equivalent power. The higher speed, however, requires the use of a gearbox and flexible couplings for main propulsion use. Cylinder sizes are smaller, requiring more units and therefore more maintenance, but the increased speed partly offsets this. Cylinder liners are of simple construction since there are no ports, but cylinder heads are more complicated and valve operating gear is required. Scavenging is a positive operation without use of scavenge trunking, thus there can be no scavenge fires. Better quality fuel is necessary because of the higher engine speed, and lubricating oil consumption is higher than for a slow-speed diesel. Engine height is reduced with trunk piston design and there are fewer moving parts per cylinder. There are, however, in total more parts for maintenance, although they are smaller and more manageable.

The Vee engine configuration is used with some medium-speed engine designs to further reduce the size and weight for a particular power.

Couplings, clutches and gearboxes

Where the shaft speed of a medium-speed diesel is not suitable for its application, e.g. where a low speed drive for a propeller is required, a gearbox must be provided. Between the engine and gearbox it is usual to fit some form of flexible coupling to dampen out vibrations. There is also often a need for a clutch to disconnect the engine from the gearbox.

Couplings

Elastic or flexible couplings allow slight misalignment and damp out or remove torque variations from the engine. The coupling may in addition function as a clutch or disconnecting device. Couplings may be mechanical, electrical, hydraulic or pneumatic in operation. It is usual to combine the function of clutch with a coupling and this is not readily possible with the mechanical coupling.

Clutches

A clutch is a device to connect or separate a driving unit from the unit it drives. With two engines connected to a gearbox a clutch enables one or both engines to be run, and facilitates reversing of the engine.

The hydraulic or fluid coupling uses oil to connect the driving section or impeller with the driven section or runner (Figure 2.25). No wear will thus take place between these two, and the clutch operates smoothly. The runner and impeller have pockets that face each other which are filled with oil as they rotate. The engine driven impeller provides kinetic energy to the oil which transmits the drive to the runner. Thrust bearings must be provided on either side of the coupling because of the axial thrust developed by this coupling.

A plate-type clutch consists of pressure plates and clutch plates arranged in a clutch spider (Figure 2.26). A forward and an aft clutch assembly are provided, and an externally mounted selector valve assembly is the control device which hydraulically engages the desired clutch. The forward clutch assembly is made up of the input shaft and the forward clutch spider. The input shaft includes the forward driven gear and, at its extreme end, a hub with the steel pressure plates of the

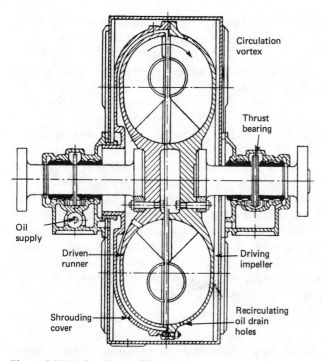

Figure 2.25 Hydraulic coupling

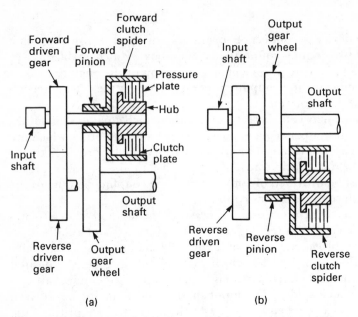

Figure 2.26 Plate-type clutch

forward clutch assembly spline-connected, i.e. free to slide. Thus when the input shaft turns, the forward driven gear and the forward clutch pressure plates will rotate. The forward clutch plates are positioned between the pressure plates and are spline-connected to the forward clutch spider or housing. This forward clutch spider forms part of the forward pinion assembly which surrounds but does not touch the input shaft. The construction of the reverse clutch spider is similar.

Both the forward and reverse pinions are in constant mesh with the output gear wheel which rotates the output shaft. In the neutral position the engine is rotating the input shaft and both driven gear wheels, but not the output shaft. When the clutch selector valve is moved to the ahead position, a piston assembly moves the clutch plates and pressure plates into contact. A friction grip is created between the smooth pressure plate and the clutch plate linings and the forward pinion rotates. The forward pinion drives the output shaft and forward propulsion will occur. The procedure when the selector valve is moved to the astern position is similar but now the reverse pinion drives the output shaft in the opposite direction.

Gearboxes

The gearing arrangement used to reduce the medium-speed engine drive down to suitable propeller revolutions is always single reduction and usually single helical. Reduction ratios range from about 2:1 to 4:1 on modern installations.

Pinion and gearwheel arrangements will be similar to those for steam turbines as described in Chapter 3, except that they will be single helical or epicyclic.

Reversing

Where a gearbox is used with a diesel engine, reversing gears may be incorporated so that the engine itself is not reversed. Where a controllable pitch propeller is in use there is no requirement to reverse the main engine. However, when it is necessary to run the engine in reverse it must be started in reverse and the fuel injection timing must be changed. Where exhaust timing or poppet valves are used they also must be retimed. With jerk-type fuel pumps the fuel cams on the camshaft must be repositioned. This can be done by having a separate reversing cam and moving the camshaft axially to bring it into position. Alternatively a lost-motion clutch may be used in conjunction with the ahead pump-timing cam.

The fuel pump cam and lost-motion clutch arrangement is shown in Figure 2.27. The shaping of the cam results in a period of pumping first then about 10° of fuel injection before top dead centre and about 5° after top dead centre. A period of dwell then occurs when the fuel pump plunger does not move. A fully reversible cam will be symmetrical about this point, as shown. The angular period between the top dead centre points for ahead and astern running will be the 'lost motion' required for astern running. The lost-motion clutch or servo motor uses a rotating vane which is attached to the camshaft but can move in relation to the camshaft drive from the crankshaft. The vane is shown held in the ahead operating position by oil pressure. When oil is supplied under pressure through the drain, the vane will rotate through the lost-motion angular distance to change the fuel timing for astern operation. The starting air system is retimed, either by this camshaft movement or by a directional air supply being admitted to the starting air distributor, to reposition the cams. Exhaust timing or poppet valves will have their own lost-motion clutch or servo motor for astern timing.

Some typical marine diesel engines

Sulzer

The RTA72 is a single-acting, low-speed, two-stroke reversible marine diesel engine manufactured by Sulzer Brothers Ltd. It is one of the RTA series engines which were introduced in 1981 and in addition to a

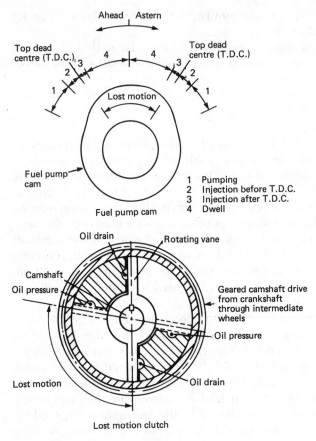

Figure 2.27 Reversing arrangements

longer stroke than the earlier RL series, it has a cylinder-head exhaust valve providing uniflow scavenging.

The bedplate is single-walled and arranged with an integral thrust bearing housing at the aft end (Figure 2.28). Cross members are steel fabrications although the centre section, incorporating the main bearing saddle tie-bolt housings, may be a steel forging. To resist crankshaft loading and transverse bending, the main bearing keeps are held down by jackbolts.

The crankcase chamber is arranged by using individual A-frames for columns which are also the mountings for the double-slippered crosshead guides. The A-frames are joined together by heavy steel plates and short angle girders to form a sturdy box frame. The A-frames in way of the thrust block are manufactured as a one-piece double column

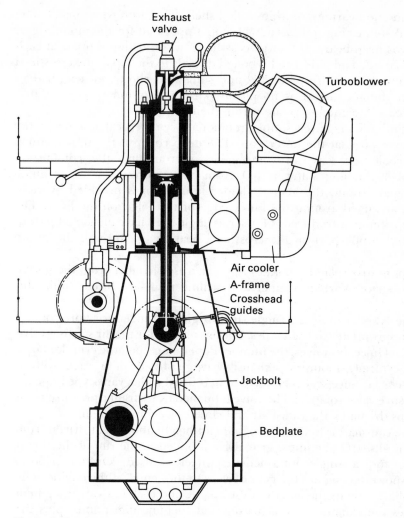

Figure 2.28 Sulzer RTA72 engine

to ensure accurate mesh of the camshaft drive gears which are enclosed in this section.

Individual cast-iron cylinder blocks are bolted together to form a rigid unit which is mounted onto the A-frames. Tie bolts extend from the top of the cylinder block to the underside of the main bearing saddles.

The crankshaft is semi-built, with the combined crankpin and crankweb elements forged from a single element. The journal pins are then shrunk into the crankwebs. For all but the larger numbers of engine cylinders, the crankshaft is a single unit. The main journal and bottom-end bearings are thin-walled shells lined with white metal. The forged connecting rod has a 'table top' upper end for the mounting of the crosshead bearing. A large crosshead, with floating slippers at each end, is used. The piston rod is bolted directly to the top of the crosshead pin. The pistons are oil-cooled and somewhat shorter in length than earlier designs. There is no piston skirt. Five piston rings are fitted which are designed to rotate within their grooves.

Cylinder liners have a simple, rotationally symmetrical design with the scavenge ports at the lower end. The deep collar at the upper end is bore-cooled, as are all components surrounding the combustion chamber. Cooling water is fed from below through a water guide arranged around the liner. Cylinder lubrication is provided by eight quills arranged around the lower edge of the collar on the liner. The more recently introduced RTA series engines all have oil-cooled pistons with oil supplied from the crosshead bearing up through the piston rod.

A piston rod gland separates the crankcase chamber from the under piston space. Various scraper and sealing rings are fitted within the gland.

The cylinder head is a single steel forging arranged for bore cooling with appropriately drilled holes. Pockets are cut for the air starting valve and fuel injection valves, the number depending upon the cylinder bore. The centrally mounted exhaust valve is fitted in a cage with a bore-cooled valve seat. The valve stem is fitted with a vane-type impeller to ensure valve rotation. The valve is opened by hydraulic pressure from pumps driven by the camshaft and closed by compressed air.

The camshaft is located at engine mid-height and is gear driven from the crankshaft. The initial gear drive is bolted to the rim of the thrust block and a single intermediate wheel is used. On larger-bore, high-powered engines the gear drive is in the centre of the engine. The camshaft extends the length of the engine and each individual segment carries the exhaust valve actuating and fuel-injection pumps plus the reversing servo motor for one pair of cylinders.

Constant-pressure turbocharging is used, and electrically driven blowers cut in automatically when the engine load is at about 40% of the maximum continuous rating.

Lubricating oil is supplied to a low- and a medium-pressure system. The low-pressure system supplies the main and other bearings. The crosshead bearing, reversing servo motors and exhaust valve actuators are supplied by the medium-pressure system. Cylinder oil is supplied to lubricators from a high-level service tank.

Fuel injection uses the jerk pump system and a Woodward-type hydraulic governor is used to control engine speed. Where the engine has oil-cooled pistons they will be supplied from the lubricating oil system, possibly at a higher pressure produced by booster pumps, e.g. the Sulzer RTA engine. An appropriate type of lubricating oil must be used for oil-lubricated pistons in order to avoid carbon deposits on the hotter parts of the system.

MAN B&W

The L70MC is a single-acting, low-speed two-stroke reversible marine diesel engine manufactured by MAN B&W. It is one of the MC series introduced in 1982, and has a longer stroke and increased maximum pressure when compared with the earlier L-GF and L-GB designs.

The bedplate is made of welded longitudinal girders and welded cross girders with cast-steel bearing supports (Figure 2.29). The frame box is mounted on the bedplate and may be of cast or welded design. On the exhaust side of the engine a relief valve and manhole are provided for each cylinder. On the camshaft side a larger hinged door is provided. The cylinder frame units which comprise one or more cylinders are of cast iron and bolted together to form the requisite number of engine cylinders. Together with the cylinder liners they form the scavenge air space and the cooling water space. The double bottom in the scavenge space is water cooled. The stuffing box fitted around the piston rod has sealing rings to stop the leakage of scavenge air and scraper rings to prevent oil entering the scavenge space.

On the camshaft side, access covers are provided for inspection and cleaning of the scavenge space. The cylinder cover is a single piece of forged steel, and has bored holes for cooling water circulation. It has a central opening for the exhaust valves and appropriate pockets for the fuel valves, a relief valve, a starting air valve and the indicator cock. The exhaust valve housing is fitted into the centre of the cylinder head. It is opened hydraulically and closed by air pressure. During operation the exhaust valve rotates. The bedplate, frame box and cylinder frames are connected together with staybolts to form the individual units. Each staybolt is braced to prevent transverse oscillations.

The crankshaft may be solid or semi-built on a cylinder by cylinder basis. A shaft piece with a thrust collar is incorporated into the crankshaft and at the after end has a flange for the turning wheel. At the forward end a flange is fitted for the mounting of a tuning device or counterweights.

The running gear consists of a piston, a piston rod and crosshead assembly and a forged steel connecting rod. The crosshead moves in guide shoes which are fitted on the frame box ends. The camshaft has several sections, each of which consists of a shaft piece with exhaust cams, fuel cams and couplings. It is driven by a chain drive from the crankshaft.

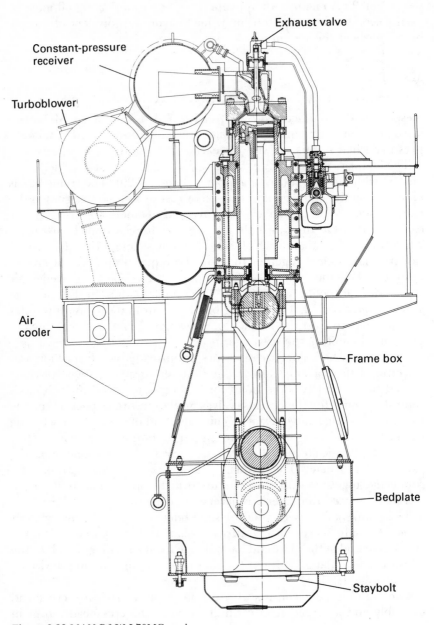

Figure 2.29 MAN B&W L70MC engine

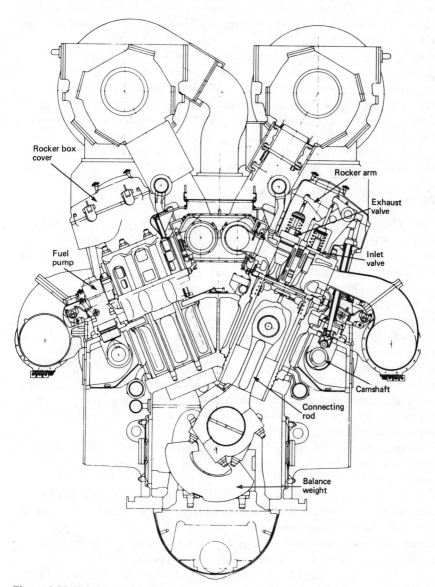

Rocker box
cover

Rocker arm

Exhaust
valve

Fuel
pump

Inlet
valve

Camshaft

Connecting
rod

Balance
weight

Figure 2.30 Pielstick PC4 engine

Exhaust gas from the engine is passed into a constant-pressure receiver and then into the turbochargers. Scavenging is uniflow, and electrically driven auxiliary blowers are automatically started during low-load operation.

Lubricating oil is supplied to the various bearings and also to the pistons for cooling. Cylinder oil is supplied via lubricators from a high-level service tank. A separate lubrication system is provided for the camshaft bearings to prevent contamination of the main lubricating oil system. Fresh water cooling is provided for the cylinder jackets, cylinder covers and exhaust valves.

The engine is designed to run on diesel oil or heavy fuel oil. An electronic governor is provided as standard.

Pielstick

The Pielstick PC series engines are single-acting, medium-speed, four-stroke reversible types. Both in-line and V-configurations are available. The running gear, being a trunk-type engine, is made up of the piston and the connecting rod which joins the single-throw crankshaft. The arrangement of a PC4 engine is shown in Figure 2.30.

The crankcase and frame are constructed from heavy plate and steel castings to produce a low-weight rigid structure. The crankshaft is underslung and this arrangement confines all stresses to the frame structure. The crankshaft is a one-piece forging and the connecting rods are H-section steel stampings. The one-piece cylinder head contains two exhaust and two inlet valves together with a starting air valve, a relief valve, indicator cock and a centrally positioned fuel injector.

Exhaust-gas-driven turbo-chargers operating on the pulse system supply pressurised air to the engine cylinders.

Bearing lubrication and piston cooling are supplied from a common system. The engine has a dry sump with oil suction being taken from a separate tank.

The cylinder jackets are water-cooled together with the cylinder heads and the exhaust valve cages. The charge air cooler may be fresh-water or sea-water cooled as required.

Fuel injection uses the jerk pump system, and a Woodward-type hydraulic governor is used to control engine speed.

Later versions of the PC series engine are described as PC20 and PC40 and have somewhat increased scantlings.

Operating procedures

Medium- and slow-speed diesel engines will follow a fairly similar procedure for starting and manœuvring. Where reversing gearboxes or controllable-pitch propellers are used then engine reversing is not necessary. A general procedure is now given for engine operation which details the main points in their correct sequence. Where a manufacturer's instruction book is available this should be consulted and used.

Preparations for standby

1. Before a large diesel is started it must be warmed through by circulating hot water through the jackets, etc. This will enable the various engine parts to expand in relation to one another.
2. The various supply tanks, filters, valves and drains are all to be checked.
3. The lubricating oil pumps and circulating water pumps are started and all the visible returns should be observed.
4. All control equipment and alarms should be examined for correct operation.
5. The indicator cocks are opened, the turning gear engaged and the engine turned through several complete revolutions. In this way any water which may have collected in the cylinders will be forced out.
6. The fuel oil system is checked and circulated with hot oil.
7. Auxiliary scavenge blowers, if manually operated, should be started.
8. The turning gear is removed and if possible the engine should be turned over on air before closing the indicator cocks.
9. The engine is now available for standby.

The length of time involved in these preparations will depend upon the size of the engine.

Engine starting

1. The direction handle is positioned ahead or astern. This handle may be built into the telegraph reply lever. The camshaft is thus positioned relative to the crankshaft to operate the various cams for fuel injection, valve operation, etc.
2. The manœuvring handle is moved to 'start'. This will admit compressed air into the cylinders in the correct sequence to turn the engine in the desired direction.A separate air start button may be used.
3. When the engine reaches its firing speed the manœuvring handle is moved to the running position. Fuel is admitted and the combustion process will accelerate the engine and starting air admission will cease.

Engine reversing

When running at manœuvring speeds:

1. Where manually operated auxiliary blowers are fitted they should be started.
2. The fuel supply is shut off and the engine will quickly slow down.
3. The direction handle is positioned astern.
4. Compressed air is admitted to the engine to turn it in the astern direction.
5. When turning astern under the action of compressed air, fuel will be admitted. The combustion process will take over and air admission cease.

When running at full speed:

1. The auxiliary blowers, where manually operated, should be started.
2. Fuel is shut off from the engine.
3. Blasts of compressed air may be used to slow the engine down.
4. When the engine is stopped the direction handle is positioned astern.
5. Compressed air is admitted to turn the engine astern and fuel is admitted to accelerate the engine. The compressed air supply will then cease.

Chapter 3
Steam turbines and gearing

The steam turbine has until recently been the first choice for very large power main propulsion units. Its advantages of little or no vibration, low weight, minimal space requirements and low maintenance costs are considerable. Furthermore a turbine can be provided for any power rating likely to be required for marine propulsion. However, the higher specific fuel consumption when compared with a diesel engine offsets these advantages, although refinements such as reheat have narrowed the gap.

The steam turbine is a device for obtaining mechanical work from the energy stored in steam. Steam enters the turbine with a high energy content and leaves after giving up most of it. The high-pressure steam from the boiler is expanded in nozzles to create a high-velocity jet of steam. The nozzle acts to convert heat energy in the steam into kinetic energy. This jet is directed into blades mounted on the periphery of a wheel or disc (Figure 3.1). The steam does not 'blow the wheel around'. The shaping of the blades causes a change in direction and hence velocity of the steam jet. Now a change in velocity for a given mass flow of steam will produce a force which acts to turn the turbine wheel, i.e. mass flow of steam (kg/s) × change in velocity (m/s) = force (kg m/s^2).

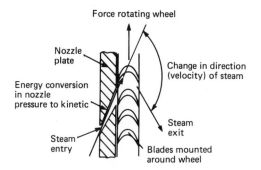

Figure 3.1 Energy conversion in a steam turbine

This is the operating principle of all steam turbines, although the arrangements may vary considerably. The steam from the first set of blades then passes to another set of nozzles and then blades and so on along the rotor shaft until it is finally exhausted. Each set comprising nozzle and blades is called a stage.

Turbine types

There are two main types of turbine, the 'impulse' and the 'reaction'. The names refer to the type of force which acts on the blades to turn the turbine wheel.

Impulse

The impulse arrangement is made up of a ring of nozzles followed by a ring of blades. The high-pressure, high-energy steam is expanded in the nozzle to a lower-pressure, high-velocity jet of steam. This jet of steam is directed into the impulse blades and leaves in a different direction (Figure 3.2). The changing direction and therefore velocity produces an impulsive force which mainly acts in the direction of rotation of the turbine blades. There is only a very small end thrust on the turbine shaft.

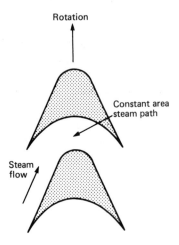

Figure 3.2 Impulse blading

Reaction

The reaction arrangement is made up of a ring of fixed blades attached to the casing, and a row of similar blades mounted on the rotor, i.e.

moving blades (Figure 3.3). The blades are mounted and shaped to produce a narrowing passage which, like a nozzle, increases the steam velocity. This increase in velocity over the blade produces a reaction force which has components in the direction of blade rotation and also along the turbine axis. There is also a change in velocity of the steam as a result of a change in direction and an impulsive force is also produced with this type of blading. The more correct term for this blade arrangement is 'impulse-reaction'.

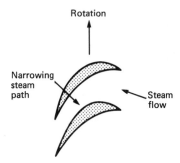

Figure 3.3 Reaction blading

Compounding

Compounding is the splitting up, into two or more stages, of the steam pressure or velocity change through a turbine.

Pressure compounding of an impulse turbine is the use of a number of stages of nozzle and blade to reduce progressively the steam pressure. This results in lower or more acceptable steam flow speeds and a better turbine efficiency.

Velocity compounding of an impulse turbine is the use of a single nozzle with an arrangement of several moving blades on a single disc. Between the moving blades are fitted guide blades which are connected to the turbine casing. This arrangement produces a short lightweight turbine with a poorer efficiency which would be acceptable in, for example, an astern turbine.

The two arrangements may be combined to give what is called 'pressure-velocity compounding'.

The reaction turbine as a result of its blade arrangement changes the steam velocity in both fixed and moving blades with consequent gradual steam pressure reduction. Its basic arrangement therefore provides compounding.

The term 'cross-compound' is used to describe a steam turbine unit made up of a high pressure and a low pressure turbine (Figure 3.4). This is the usual main propulsion turbine arrangement. The alternative is a

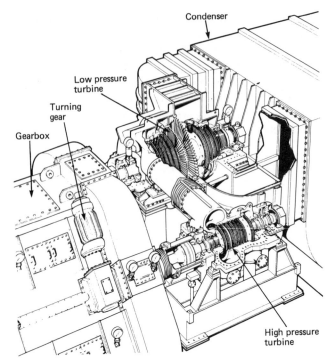

Figure 3.4 Cross compound turbine arrangement

single cylinder unit which would be usual for turbo-generator sets, although some have been fitted for main propulsion service.

Reheat

Reheating is a means of improving the thermal efficiency of the complete turbine plant. Steam, after expansion in the high-pressure turbine, is returned to the boiler to be reheated to the original superheat temperature. It is then returned to the turbine and further expanded through any remaining stages of the high-pressure turbine and then the low-pressure turbine.

Named turbine types

A number of famous names are associated with certain turbine types.

Parsons. A reaction turbine where steam expansion takes place in the fixed and moving blades. A stage is made up of one of each blade type. Half of the stage heat drop occurs in each blade type, therefore providing 50% reaction per stage.

Curtis. An impulse turbine with more than one row of blades to each row of nozzles, i.e. velocity compounded.

De Laval. A high-speed impulse turbine which has only one row of nozzles and one row of blades.

Rateau. An impulse turbine with several stages, each stage being a row of nozzles and a row of blades, i.e. pressure compounded.

Astern arrangements

Marine steam turbines are required to be reversible. This is normally achieved by the use of several rows of astern blading fitted to the high-pressure and low-pressure turbine shafts to produce astern turbines. About 50% of full power is achieved using these astern turbines. When the turbine is operating ahead the astern blading acts as an air compressor, resulting in windage and friction losses.

Turbine construction

The construction of an impulse turbine is shown in Figure 3.5. The turbine rotor carries the various wheels around which are mounted the blades. The steam decreases in pressure as it passes along the shaft and increases in volume requiring progressively larger blades on the wheels. The astern turbine is mounted on one end of the rotor and is much

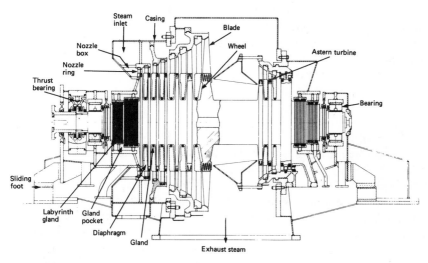

Figure 3.5 Impulse turbine

shorter than the ahead turbine. The turbine rotor is supported by bearings at either end; one bearing incorporates a thrust collar to resist any axial loading.

The turbine casing completely surrounds the rotor and provides the inlet and exhaust passages for the steam. At the inlet point a nozzle box is provided which by use of a number of nozzle valves admits varying amounts of steam to the nozzles in order to control the power developed by the turbine. The first set of nozzles is mounted in a nozzle ring fitted into the casing. Diaphragms are circular plates fastened to the casing which are fitted between the turbine wheels. They have a central circular hole through which the rotor shaft passes. The diaphragms contain the nozzles for steam expansion and a gland is fitted between the rotor and the diaphragm.

The construction of a reaction turbine differs somewhat in that there are no diaphragms fitted and instead fixed blades are located between the moving blades.

Rotor

The turbine rotor acts as the shaft which transmits the mechanical power produced to the propeller shaft via the gearing. It may be a single piece with the wheels integral with the shaft or built up from a shaft and separate wheels where the dimensions are large.

The rotor ends adjacent to the turbine wheels have an arrangement of raised rings which form part of the labyrinth gland sealing system, described later in this chapter. Journal bearings are fitted at each end of the rotor. These have rings arranged to stop oil travelling along the shaft which would mix with the steam. One end of the rotor has a small thrust collar for correct longitudinal alignment. The other end has an appropriate flange or fitting arranged for the flexible coupling which joins the rotor to the gearbox pinion.

The blades are fitted into grooves of various designs cut into the wheels.

Blades

The shaping and types of turbine blades have already been discussed. When the turbine rotor is rotating at high speed the blades will be subjected to considerable centrifugal force and variations in steam velocity across the blades will result in blade vibration.

Expansion and contraction will also occur during turbine operation, therefore a means of firmly securing the blades to the wheel is essential. A number of different designs have been employed (Figure 3.6).

Fitting the blades involves placing the blade root into the wheel

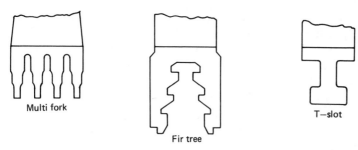

Multi fork

Fir tree

T—slot

Figure 3.6 Blade fastening

through a gate or entrance slot and sliding it into position. Successive blades are fitted in turn and the gate finally closed with a packing piece which is pinned into place. Shrouding is then fitted over tenons on the upper edge of the blades. Alternatively, lacing wires may be passed through and brazed to all the blades.

End thrust

In a reaction turbine a considerable axial thrust is developed. The closeness of moving parts in a high-speed turbine does not permit any axial movement to take place: the axial force or end thrust must therefore be balanced out.

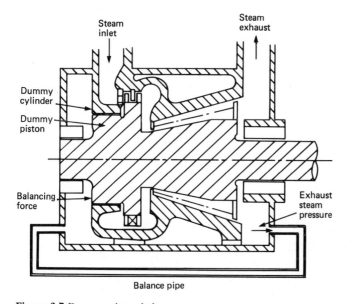

Figure 3.7 Dummy piston balance arrangement

One method of achieving this balance is the use of a dummy piston and cylinder. A pipe from some stage in the turbine provides steam to act on the dummy piston which is mounted on the turbine rotor (Figure 3.7). The rotor casing provides the cylinder to enable the steam pressure to create an axial force on the turbine shaft. The dummy piston annular area and the steam pressure are chosen to produce a force which exactly balances the end thrust from the reaction blading. A turbine with ahead and astern blading will have a dummy piston at either end to ensure balance in either direction of rotation.

Another method often used in low-pressure turbines is to make the turbine double flow. With this arrangement steam enters at the centre of the shaft and flows along in opposite directions. With an equal division of steam the two reaction effects balance and cancel one another.

Glands and gland sealing

Steam is prevented from leaking out of the rotor high-pressure end and air is prevented from entering the low-pressure end by the use of glands. A combination of mechanical glands and a gland sealing system is usual.

Mechanical glands are usually of the labyrinth type. A series of rings projecting from the rotor and the casing combine to produce a maze of winding passages or a labyrinth (Figure 3.8). Any escaping steam must pass through this labyrinth, which reduces its pressure progressively to zero.

The gland sealing system operates in conjunction with the labyrinth gland where a number of pockets are provided. The system operates in one of two ways.

When the turbine is running at full speed steam will leak into the first pocket and a positive pressure will be maintained there. Any steam which further leaks along the shaft to the second pocket will be extracted

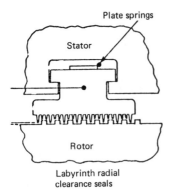

Figure 3.8 Labyrinth glands

by an air pump or air ejector to the gland steam condenser. Any air which leaks in from the machinery space will also pass to the gland steam condenser (Figure 3.9).

At very low speeds or when starting up, steam is provided from a low-pressure supply to the inner pocket. The outer pocket operates as before.

The gland steam sealing system provides the various low-pressure steam supplies and extraction arrangements for all the glands in the turbine unit.

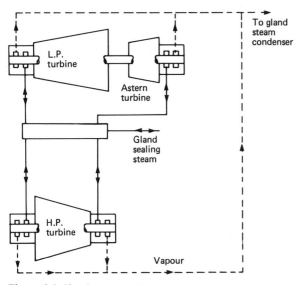

Figure 3.9 Gland steam sealing system

Diaphragms

Only impulse turbines have diaphragms. Diaphragms are circular plates made up of two semi-circular halves. A central semi-circular hole in each is provided for the shaft to pass through. The diaphragm fits between the rotor wheels and is fastened into the casing. The nozzles are housed in the diaphragm around its periphery. The central hole in the diaphragm is arranged with projections to produce a labyrinth gland around the shaft.

Nozzles

Nozzles serve to convert the high pressure and high energy of the steam into a high-velocity jet of steam with a reduced pressure and energy content. The steam inlet nozzles are arranged in several groups with all but the main group having control valves (Figure 3.10). In this way the power produced by the turbine can be varied, depending upon how many nozzle control valves are opened. Both impulse and reaction turbines have steam inlet nozzles.

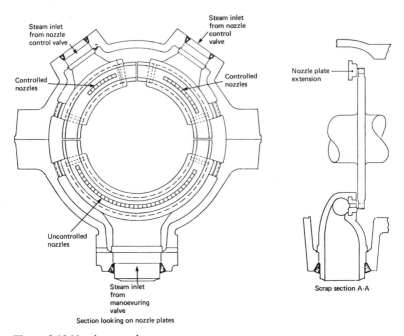

Figure 3.10 Nozzle control

Drains

During warming through operations or when manœuvring, steam will condense and collect in various places within the turbine and its pipelines. A system of drains must be provided to clear this water away to avoid its being carried over into the blades, which may do damage. Localised cooling or distortion due to uneven heating could also be caused.

Modern installations now have automatic drain valves which open when warming through or manœuvring and close when running at normal speed.

Bearings

Turbine bearings are steel backed, white-metal lined and supported in adjustable housings to allow alignment changes if required. Thrust bearings are of the tilting pad type and are spherically seated. The pads are thus maintained parallel to the collar and equally loaded. Details of both types can be seen in Figure 3.5.

Lubricating oil enters a turbine bearing through a port on either side. The entry point for the oil is chamfered to help distribute the oil along the bearing. No oil ways are provided in turbine bearings and a greater clearance between bearing and shaft is provided compared with a diesel engine. The shaft is able to 'float' on a wedge of lubricating oil during turbine operation. The oil leaves the bearing at the top and returns to the drain tank.

Lubricating oil system

Lubricating oil serves two functions in a steam turbine:

1. It provides an oil film to reduce friction between moving parts.
2. It removes heat generated in the bearings or conducted along the shaft.

A common lubrication system is used to supply oil to the turbine, gearbox and thrust bearings and the gear sprayers. The turbine, rotating at high speed, requires a considerable time to stop. If the main motor driven lubricating oil pumps were to fail an emergency supply of

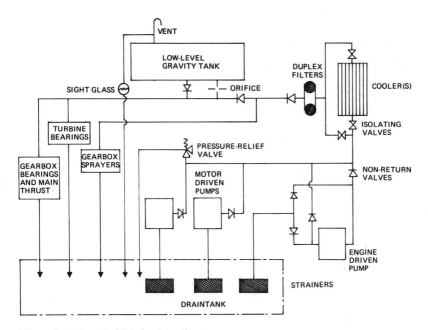

Figure 3.11 A typical lubricating oil system

lubricating oil would be necessary. This is usually provided from a gravity tank, although main engine driven lubricating oil pumps may also be required.

A lubricating oil system employing both a gravity tank and an engine driven pump is shown in Figure 3.11. Oil is drawn from the drain tank through strainers and pumped to the coolers. Leaving the coolers, the oil passes through another set of filters before being distributed to the gearbox, the turbine bearings and the gearbox sprayers. Some of the oil also passes through an orifice plate and into the gravity tank from which it continuously overflows (this can be observed through the sightglass). The engine driven pump supplies a proportion of the system requirements in normal operation.

In the event of a power failure the gearbox sprayers are supplied from the engine driven pump. The gravity tank provides a low-pressure supply to the bearings over a considerable period to enable the turbine to be brought safely to rest.

Expansion arrangements

The variation in temperature for a turbine between stationary and normal operation is considerable. Arrangements must therefore be made to permit the rotor and casing to expand.

The turbine casing is usually fixed at the after end to a pedestal support or brackets from the gearbox. The support foot or palm on the casing is held securely against fore and aft movement, but because of elongated bolt holes may move sideways. The forward support palm has similar elongated holes and may rest on a sliding foot or panting plates. Panting plates are vertical plates which can flex or move axially as expansion takes place.

The forward pedestal and the gearcase brackets or after pedestal supports for the casing are fixed in relation to one another. The use of large vertical keys and slots on the supports and casing respectively, ensures that the casing is kept central and in axial alignment.

The rotor is usually fixed at its forward end by the thrust collar, and any axial movement must therefore be taken up at the gearbox end. Between the turbine rotor and the gearbox is fitted a flexible coupling. This flexible coupling is able to take up all axial movement of the rotor as well as correct for any slight misalignment.

Any pipes connected to the turbine casing must have large radiused bends or be fitted with bellows pieces to enable the casing to move freely. Also, any movement of the pipes due to expansion must not affect the casing. This is usually ensured by the use of flexible or spring supports on the pipes.

When warming through a turbine it is important to ensure that expansion is taking place freely. Various indicators are provided to enable this to be readily checked. Any sliding arrangements should be kept clean and well lubricated.

Turbine control

The valves which admit steam to the ahead or astern turbines are known as 'manœuvring valves'. There are basically three valves, *the ahead, the astern* and *the guarding* or *guardian* valve. The guardian valve is an astern steam isolating valve. These valves are hydraulically operated by an independent system employing a main and standby set of pumps. Provision is also made for hand operation in the event of remote control system failure.

Operation of the ahead manœuvring valve will admit steam to the main nozzle box. Remotely operated valves are used to open up the remaining nozzle boxes for steam admission as increased power is required. A speed-sensitive control device acts on the ahead manœuvring valve to hold the turbine speed constant at the desired value.

Operation of the astern manœuvring valve will admit steam to the guardian valve which is opened in conjunction with the astern valve. Steam is then admitted to the astern turbines.

Turbine protection

A turbine protection system is provided with all installations to prevent damage resulting from an internal turbine fault or the malfunction of some associated equipment. Arrangements are made in the system to shut the turbine down using an emergency stop and solenoid valve. Operation of this device cuts off the hydraulic oil supply to the manœuvring valve and thus shuts off steam to the turbine. This main trip relay is operated by a number of main fault conditions which are:

1. Low lubricating oil pressure.
2. Overspeed.
3. Low condenser vacuum.
4. Emergency stop.
5. High condensate level in condenser.
6. High or low boiler water level.

Other fault conditions which must be monitored and form part of a total protection system are:

1. HP and LP rotor eccentricity or vibration.
2. HP and LP turbine differential expansion, i.e. rotor with respect to casing.

3. HP and LP thrust bearing weardown.
4. Main thrust bearing weardown.
5. Turning gear engaged (this would prevent starting of the turbine).

Such 'turbovisory' systems, as they may be called, operate in two ways. If a tendency towards a dangerous condition is detected a first stage alarm is given. This will enable corrective action to be taken and the turbine is not shut down. If corrective action is not rapid, is unsuccessful, or a main fault condition quickly arises, the second stage alarm is given and the main trip relay is operated to stop the turbine.

Gearing

Steam turbines operate at speeds up to 6000 rev/min. Medium-speed diesel engines operate up to about 750 rev/min. The best propeller speed for efficient operation is in the region of 80 to 100 rev/min. The turbine or engine shaft speed is reduced to that of the propeller by the use of a system of gearing. Helical gears have been used for many years and remain a part of most systems of gearing. Epicyclic gears with their compact, lightweight, construction are being increasingly used in marine transmissions.

Epicyclic gearing

This is a system of gears where one or more wheels travel around the outside or inside of another wheel whose axis is fixed. The different arrangements are known as planetary gear, solar gear and star gear (Figure 3.12).

The wheel on the principal axis is called the sun wheel. The wheel whose centre revolves around the principal axis is the planet wheel. An internal-teeth gear which meshes with the planet wheel is called the annulus. The different arrangements of fixed arms and sizing of the sun and planet wheels provide a variety of different reduction ratios.

Steam turbine gearing may be double or triple reduction and will be a combination from input to output of star and planetary modes in conjunction with helical gearing (Figure 3.13).

Helical gearing

Single or double reduction systems may be used, although double reduction is more usual. With single reduction the turbine drives a pinion with a small number of teeth and this pinion drives the main wheel which is directly coupled to the propeller shaft. With double

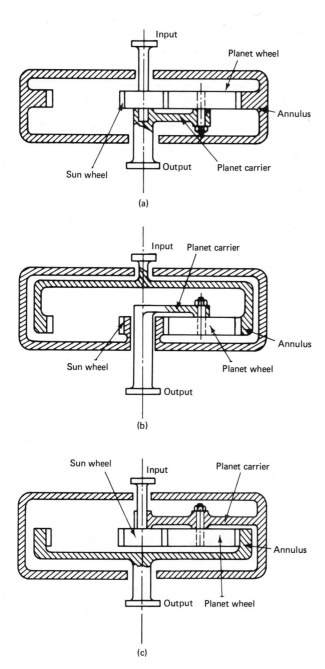

Figure 3.12 Epicyclic gearing; (a) planetary gear, (b) solar gear, (c) star gear

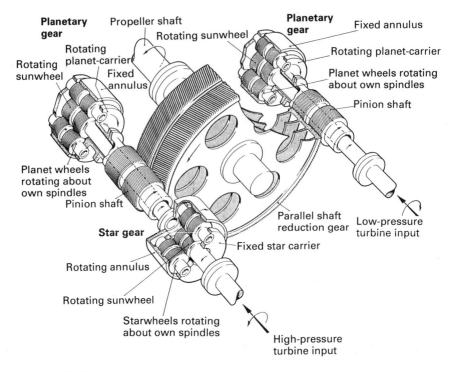

Figure 3.13 Typical marine turbine reduction gear

reduction the turbine drives a primary pinion which drives a primary wheel. The primary wheel drives, on the same shaft, a secondary pinion which drives the main wheel. The main wheel is directly coupled to the propeller shaft. A double reduction gearing system is shown in Figure 3.14.

All modern marine gearing is of the double helical type. Helical means that the teeth form part of a helix on the periphery of the pinion or gear wheel. This means that at any time several teeth are in contact and thus the spread and transfer of load is much smoother. Double helical refers to the use of two wheels or pinions on each shaft with the teeth cut in opposite directions. This is because a single set of meshing helical teeth would produce a sideways force, moving the gears out of alignment. The double set in effect balances out this sideways force. The gearing system shown in Figure 3.14 is double helical.

Lubrication of the meshing teeth is from the turbine lubricating oil supply. Sprayers are used to project oil at the meshing points both above and below and are arranged along the length of the gear wheel.

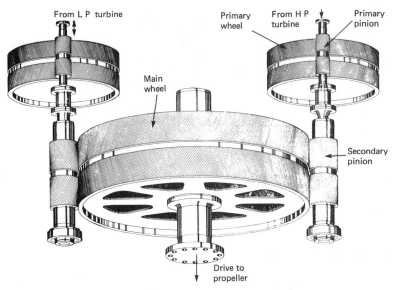

Figure 3.14 Double reduction system of gearing

Flexible coupling

A flexible coupling is always fitted between the turbine rotor and the gearbox pinion. It permits slight rotor and pinion misalignment as well as allowing for axial movement of the rotor due to expansion. Various designs of flexible coupling are in use using teeth, flexible discs, membranes, etc.

The membrane-type flexible coupling shown in Figure 3.15 is made up of a torque tube, membranes and adaptor plates. The torque tube fits between the turbine rotor and the gearbox pinion. The adaptor plates are spigoted and dowelled onto the turbine and pinion flanges and the membrane plates are bolted between the torque tube and the adaptor plates. The flexing of the membrane plates enables axial and transverse movement to take place. The torque tube enters the adaptor plate with a clearance which will provide an emergency centring should the membranes fail. The bolts in their clearance holes would provide the continuing drive until the shaft could be stopped.

Turning gear

The turning gear on a turbine installation is a reversible electric motor driving a gearwheel which meshes into the high-pressure turbine primary pinion. It is used for gearwheel and turbine rotation during maintenance or when warming-through prior to manoeuvring.

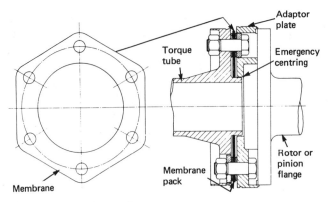

Figure 3.15 Flexible coupling

Operating procedures

The steam turbine requires a considerable period for warming-through prior to any manœuvring taking place. The high-speed operation of the turbine and its simply supported rotor also require great care during manœuvring operations.

Warming-through a steam turbine   ⊥ (c)

First open all the turbine-casing and main steam-line drain valves and ensure that all the steam control valves at the manœuvring station and around the turbine are closed. All bled steam-line drain valves should be opened. Start the lubricating oil pump and see that the oil is flowing freely to each bearing and gear sprayer, venting off air if necessary and check that the gravity tank is overflowing.

Obtain clearance from the bridge to turn the shaft. Engage the turning gear and rotate the turbines in each direction.

Start the sea water circulating pump for the main condenser. Then start the condensate extraction pump with the air ejector recirculation valve wide open. Open the manœuvring valve bypass or 'warming through' valve, if fitted. This allows a small quantity of steam to pass through the turbine and heat it. Raising a small vacuum in the condenser will assist this warming through. The turbines should be continuously turned with the turning gear until a temperature of about 75°C is reached at the LP turbine inlet after about one hour. The expansion arrangements on the turbine to allow freedom of movement should be checked.

Gland sealing steam should now be partially opened up and the vacuum increased. The turning gear should now be disengaged.

Short blasts of steam are now admitted to the turbine through the main valve to spin the propeller about one revolution. This should be repeated about every three to five minutes for a period of 15 to 30 minutes. The vacuum can now be raised to its operational value and also the gland steam pressure. The turbines are now ready for use.

While waiting for the first movements from the bridge, and between movements, the turbine must be turned ahead once every five minutes by steam blasts. If there is any delay gland steam and the vacuum should be reduced.

Manœuvring

Once warmed through, the turbine rotor must not remain stationary more than a few minutes at a time because the rotor could sag or distort, which would lead to failure, if not regularly rotated.

Astern operation involves admitting steam to the astern turbines. Where any considerable period of astern running occurs turbine temperatures, noise levels, bearings, etc., must be closely observed. The turbine manufacturer may set a time limit of about 30 minutes on continuous running astern.

Emergency astern operation

If, when travelling at full speed ahead, an order for an emergency stop or astern movement is required then safe operating procedures must be ignored.

Ahead steam is shut off, probably by the use of an emergency trip, and the astern steam valve is partly opened to admit a gradually increasing amount of steam. The turbine can thus be brought quickly to a stopped condition and if required can then be operated astern.

The stopping of the turbine or its astern operation will occur about 10 to 15 minutes before a similar state will occur for the ship. The use of emergency procedures can lead to serious damage in the turbine, gearbox or boilers.

Full away

Manœuvring revolutions are usually about 80% of the full away or full speed condition. Once the full away command is received the turbine can gradually be brought up to full power operation, a process taking one to two hours. This will also involve bringing into use turbo-alternators which use steam removed or 'bled' at some stage from the main turbines.

Checks should be made on expansion arrangements, drains should be checked to be closed, the condensate recirculation valve after the air ejector should be closed, and the astern steam valves tightly closed.

Port arrival

Prior to arriving at a port the bridge should provide one to two hours' notice to enable the turbines to be brought down to manœuvring revolutions. A diesel alternator will have to be started, the turbo-alternator shut down, and all the full away procedure done in reverse order.

Chapter 4
Boilers

A boiler in one form or another will be found on every type of ship. Where the main machinery is steam powered, one or more large water-tube boilers will be fitted to produce steam at very high temperatures and pressures. On a diesel main machinery vessel, a smaller (usually firetube type) boiler will be fitted to provide steam for the various ship services. Even within the two basic design types, watertube and firetube, a variety of designs and variations exist.

A boiler is used to heat feedwater in order to produce steam. The energy released by the burning fuel in the boiler furnace is stored (as temperature and pressure) in the steam produced. All boilers have a furnace or combustion chamber where fuel is burnt to release its energy. Air is supplied to the boiler furnace to enable combustion of the fuel to take place. A large surface area between the combustion chamber and the water enables the energy of combustion, in the form of heat, to be transferred to the water.

A drum must be provided where steam and water can separate. There must also be a variety of fittings and controls to ensure that fuel oil, air and feedwater supplies are matched to the demand for steam. Finally there must be a number of fittings or mountings which ensure the safe operation of the boiler.

In the steam generation process the feedwater enters the boiler where it is heated and becomes steam. The feedwater circulates from the steam drum to the water drum and is heated in the process. Some of the feedwater passes through tubes surrounding the furnace, i.e. waterwall and floor tubes, where it is heated and returned to the steam drum. Large-bore downcomer tubes are used to circulate feedwater between the drums. The downcomer tubes pass outside of the furnace and join the steam and water drums. The steam is produced in a steam drum and may be drawn off for use from here. It is known as 'wet' or saturated steam in this condition because it will contain small quantities of water. Alternatively the steam may pass to a superheater which is located within the boiler. Here steam is further heated and 'dried', i.e. all traces of water are converted into steam. This superheated steam then leaves the boiler for use in the system. The temperature of superheated steam will be above that of the steam in the drum. An 'attemperator', i.e. a steam cooler, may be fitted in the system to control the superheated steam temperature.

The hot gases produced in the furnace are used to heat the feedwater to produce steam and also to superheat the steam from the boiler drum. The gases then pass over an economiser through which the feedwater passes before it enters the boiler. The exhaust gases may also pass over an air heater which warms the combustion air before it enters the furnace. In this way a large proportion of the heat energy from the hot gases is used before they are exhausted from the funnel. The arrangement is shown in Figure 4.1.

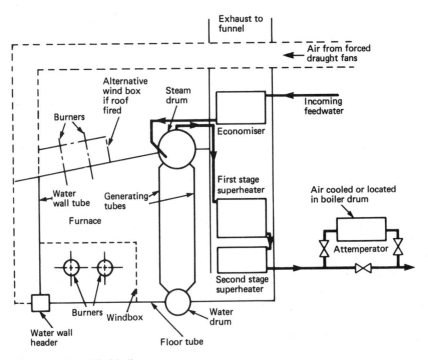

Figure 4.1 Simplified boiler arrangement

Two basically different types of boiler exist, namely the watertube and the firetube. In the watertube the feedwater is passed through the tubes and the hot gases pass over them. In the firetube boiler the hot gases pass through the tubes and the feedwater surrounds them.

Boiler types

The watertube boiler is employed for high-pressure, high-temperature, high-capacity steam applications, e.g. providing steam for main propulsion turbines or cargo pump turbines. Firetube boilers are used for auxiliary purposes to provide smaller quantities of low-pressure steam on diesel engine powered ships.

Watertube boilers

The construction of watertube boilers, which use small-diameter tubes and have a small steam drum, enables the generation or production of steam at high temperatures and pressures. The weight of the boiler is much less than an equivalent firetube boiler and the steam raising

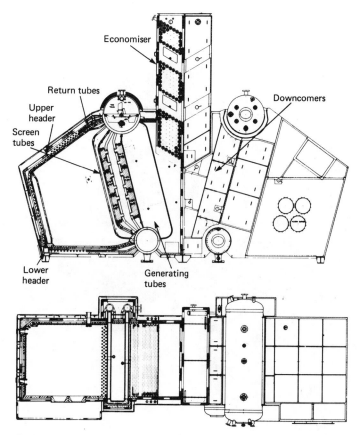

Figure 4.2 Foster Wheeler D-Type boiler

process is much quicker. Design arrangements are flexible, efficiency is high and the feedwater has a good natural circulation. These are some of the many reasons why the watertube boiler has replaced the firetube boiler as the major steam producer.

Early watertube boilers used a single drum. Headers were connected to the drum by short, bent pipes with straight tubes between the headers. The hot gases from the furnace passed over the tubes, often in a single pass.

A later development was the bent tube design. This boiler has two drums, an integral furnace and is often referred to as the 'D' type because of its shape (Figure 4.2). The furnace is at the side of the two drums and is surrounded on all sides by walls of tubes. These waterwall tubes are connected either to upper and lower headers or a lower header and the steam drum. Upper headers are connected by return tubes to the steam drum. Between the steam drum and the smaller water drum below, large numbers of smaller-diameter generating tubes are fitted.

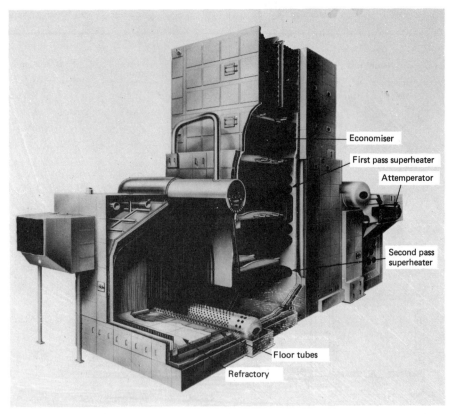

Figure 4.3 Foster Wheeler Type ESD I boiler

These provide the main heat transfer surfaces for steam generation. Large-bore pipes or downcomers are fitted between the steam and water drum to ensure good natural circulation of the water. In the arrangement shown, the superheater is located between the drums, protected from the very hot furnace gases by several rows of screen tubes. Refractory material or brickwork is used on the furnace floor, the burner wall and also behind the waterwalls. The double casing of the boiler provides a passage for the combustion air to the air control or register surrounding the burner.

The need for a wider range of superheated steam temperature control led to other boiler arrangements being used. The original External Superheater 'D' (ESD) type of boiler used a primary and secondary superheater located after the main generating tube bank (Figure 4.3). An attemperator located in the combustion air path was used to control the steam temperature.

The later ESD II type boiler was similar in construction to the ESD I but used a control unit (an additional economiser) between the primary and secondary superheaters. Linked dampers directed the hot gases over the control unit or the superheater depending upon the superheat temperature required. The control unit provided a bypass path for the gases when low temperature superheating was required.

In the ESD III boiler the burners are located in the furnace roof, which provides a long flame path and even heat transfer throughout the furnace. In the boiler shown in Figure 4.4, the furnace is fully water-cooled and of monowall construction, which is produced from finned tubes welded together to form a gastight casing. With monowall construction no refractory material is necessary in the furnace.

The furnace side, floor and roof tubes are welded into the steam and water drums. The front and rear walls are connected at either end to upper and lower water-wall headers. The lower water-wall headers are connected by external downcomers from the steam drum and the upper water-wall headers are connected to the steam drum by riser tubes.

The gases leaving the furnace pass through screen tubes which are arranged to permit flow between them. The large number of tubes results in considerable heat transfer before the gases reach the secondary superheater. The gases then flow over the primary superheater and the economiser before passing to exhaust. The dry pipe is located in the steam drum to obtain reasonably dry saturated steam from the boiler. This is then passed to the primary superheater and then to the secondary superheater. Steam temperature control is achieved by the use of an attemperator, located in the steam drum, operating between the primary and secondary superheaters.

Radiant-type boilers are a more recent development, in which the radiant heat of combustion is absorbed to raise steam, being transmitted

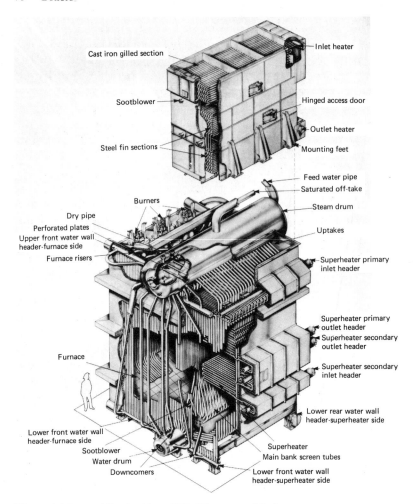

Figure 4.4 Foster Wheeler Type ESD III monowall boiler

by infra-red radiation. This usually requires roof firing and a considerable height in order to function efficiently. The ESD IV boiler shown in Figure 4.5 is of the radiant type. Both the furnace and the outer chamber are fully watercooled. There is no conventional bank of generating tubes. The hot gases leave the furnace through an opening at the lower end of the screen wall and pass to the outer chamber. The outer chamber contains the convection heating surfaces which include the primary and secondary superheaters. Superheat temperature control is by means of an attemperator in the steam drum. The hot gases, after leaving the primary superheater, pass over a steaming economiser. This is a heat exchanger in which the steam–water mixture

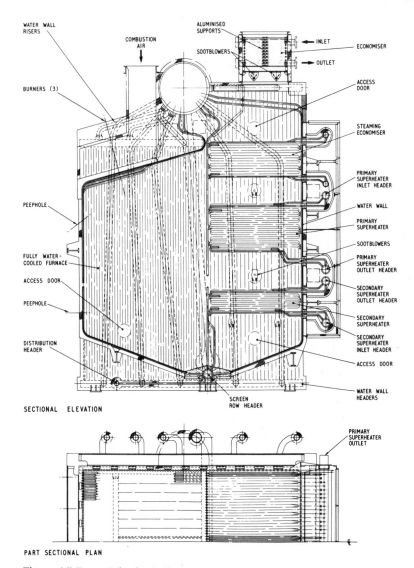

Figure 4.5 Foster Wheeler radiant-type boiler

is flowing parallel to the gas. The furnace gases finally pass over a conventional economiser on their way to the funnel.

Reheat boilers are used with reheat arranged turbine systems. Steam after expansion in the high-pressure turbine is returned to a reheater in the boiler. Here the steam energy content is raised before it is supplied to the low-pressure turbine. Reheat boilers are based on boiler designs such as the 'D' type or the radiant type.

Furnace wall construction

The problems associated with furnace refractory materials, particularly on vertical walls, have resulted in two water-wall arrangements without exposed refractory. These are known as 'tangent tube' and 'monowall' or 'membrane wall'.

In the tangent tube arrangement closely pitched tubes are backed by refractory, insulation and the boiler casing (Figure 4.6(a)). In the monowall or membrane wall arrangement the tubes have a steel strip welded between them to form a completely gas-tight enclosure (Figure 4.6(b)). Only a layer of insulation and cleading is required on the outside of this construction.

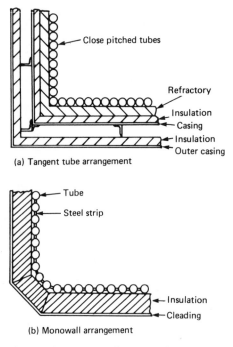

(a) Tangent tube arrangement

(b) Monowall arrangement

Figure 4.6 Furnace wall construction

The monowall construction eliminates the problems of refractory and expanded joints. However, in the event of tube failure, a welded repair must be carried out. Alternatively the tube can be plugged at either end, but refractory material must be placed over the failed tube to protect the insulation behind it. With tangent tube construction a failed tube can be plugged and the boiler operated normally without further attention.

Firetube boilers

The firetube boiler is usually chosen for low-pressure steam production on vessels requiring steam for auxiliary purposes. Operation is simple and feedwater of medium quality may be employed. The name 'tank boiler' is sometimes used for firetube boilers because of their large water capacity. The terms 'smoke tube' and 'donkey boiler' are also in use.

Package boilers

Most firetube boilers are now supplied as a completely packaged unit. This will include the oil burner, fuel pump, forced-draught fan, feed pumps and automatic controls for the system. The boiler will be fitted with all the appropriate boiler mountings.

A single-furnace three-pass design is shown in Figure 4.7. The first pass is through the partly corrugated furnace and into the cylindrical wetback combustion chamber. The second pass is back over the furnace through small-bore smoke tubes and then the flow divides at the front central smoke box. The third pass is through outer smoke tubes to the gas exit at the back of the boiler.

There is no combustion chamber refractory lining other than a lining

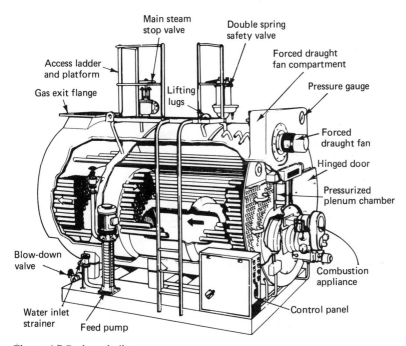

Figure 4.7 Package boiler

to the combustion chamber access door and the primary and secondary quarl.

Fully automatic controls are provided and located in a control panel at the side of the boiler.

Cochran boilers

The modern vertical Cochran boiler has a fully spherical furnace and is known as the 'spheroid' (Figure 4.8). The furnace is surrounded by water and therefore requires no refractory lining. The hot gases make a single pass through the horizontal tube bank before passing away to exhaust. The use of small-bore tubes fitted with retarders ensures better heat transfer and cleaner tubes as a result of the turbulent gas flow.

Composite boilers

A composite boiler arrangement permits steam generation either by oil firing when necessary or by using the engine exhaust gases when the ship is at sea. Composite boilers are based on firetube boiler designs. The Cochran boiler, for example, would have a section of the tube bank separately arranged for the engine exhaust gases to pass through and exit via their own exhaust duct.

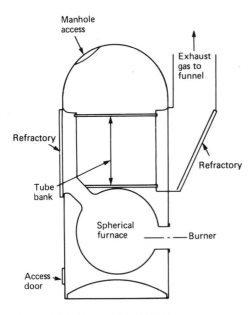

Figure 4.8 Cochran spheroid boiler

Other boiler arrangements

Apart from straightforward watertube and firetube boilers, other steam raising equipment is in use, e.g. the steam-to-steam generator, the double evaporation boiler and various exhaust gas boiler arrangements.

The steam-to-steam generator

Steam-to-steam generators produce low-pressure saturated steam for domestic and other services. They are used in conjunction with watertube boilers to provide a secondary steam circuit which avoids any possible contamination of the primary-circuit feedwater. The arrangement may be horizontal or vertical with coils within the shell which heat the feedwater. The coils are supplied with high-pressure, high-temperature steam from the main boiler. A horizontal steam-to-steam generator is shown in Figure 4.9.

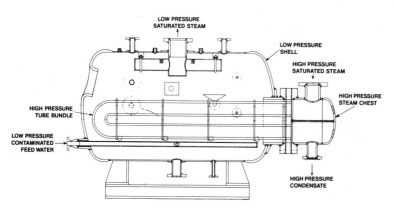

Figure 4.9 Steam-to-steam generator

Double evaporation boilers

A double evaporation boiler uses two independent systems for steam generation and therefore avoids any contamination between the primary and secondary feedwater. The primary circuit is in effect a conventional watertube boiler which provides steam to the heating coils of a steam-to-steam generator, which is the secondary system. The complete boiler is enclosed in a pressurised casing.

Exhaust gas heat exchangers

The use of exhaust gases from diesel main propulsion engines to generate steam is a means of heat energy recovery and improved plant efficiency.

An exhaust gas heat exchanger is shown in Figure 4.10. It is simply a row of tube banks circulated by feedwater over which the exhaust gases flow. Individual banks may be arranged to provide feed heating, steam generation and superheating. A boiler drum is required for steam generation and separation to take place and use is usually made of the drum of an auxiliary boiler.

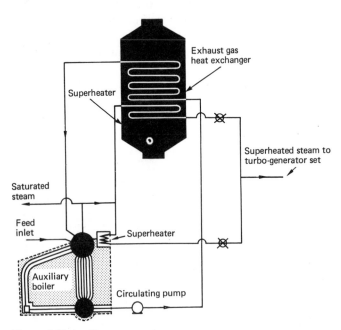

Figure 4.10 Auxiliary steam plant system

Auxiliary steam plant system

The auxiliary steam installation provided in modern diesel powered tankers usually uses an exhaust gas heat exchanger at the base of the funnel and one or perhaps two watertube boilers (Figure 4.10). Saturated or superheated steam may be obtained from the auxiliary boiler. At sea it acts as a steam receiver for the exhaust-gas heat exchanger, which is circulated through it. In port it is oil-fired in the usual way.

Exhaust gas boilers

Auxiliary boilers on diesel main propulsion ships, other than tankers, are usually of composite form, enabling steam generation using oil firing or the exhaust gases from the diesel engine. With this arrangement the boiler acts as the heat exchanger and raises steam in its own drum.

Boiler mountings

Certain fittings are necessary on a boiler to ensure its safe operation. They are usually referred to as boiler mountings. The mountings usually found on a boiler are:

Safety valves. These are mounted in pairs to protect the boiler against overpressure. Once the valve lifting pressure is set in the presence of a Surveyor it is locked and cannot be changed. The valve is arranged to open automatically at the pre-set blow-off pressure.

Main steam stop valve. This valve is fitted in the main steam supply line and is usually of the non-return type.

Auxiliary steam stop valve. This is a smaller valve fitted in the auxiliary steam supply line, and is usually of the non-return type.

Feed check or control valve. A pair of valves are fitted: one is the main valve, the other the auxiliary or standby. They are non-return valves and must give an indication of their open and closed position.

Water level gauge. Water level gauges or 'gauge glasses' are fitted in pairs, at opposite ends of the boiler. The construction of the level gauge depends upon the boiler pressure.

Pressure gauge connection. Where necessary on the boiler drum, superheater, etc., pressure gauges are fitted to provide pressure readings.

Air release cock. These are fitted in the headers, boiler drum, etc., to release air when filling the boiler or initially raising steam.

Sampling connection. A water outlet cock and cooling arrangement is provided for the sampling and analysis of feed water. A provision may also be made for injecting water treatment chemicals.

Blow down valve. This valve enables water to be blown down or emptied from the boiler. It may be used when partially or completely emptying the boiler.

Scum valve. A shallow dish positioned at the normal water level is connected to the scum valve. This enables the blowing down or removal of scum and impurities from the water surface.

Whistle stop valve. This is a small bore non-return valve which supplies the whistle with steam straight from the boiler drum.

Boiler mountings (water-tube boilers)

Watertube boilers, because of their smaller water content in relation to their steam raising capacity, require certain additional mountings:

Automatic feed water regulator. Fitted in the feed line prior to the main check valve, this device is essential to ensure the correct water level in the boiler during all load conditions. Boilers with a high evaporation rate will use a multiple-element feed water control system (see Chapter 15).

Low level alarm. A device to provide audible warning of low water level conditions.

Superheater circulating valves. Acting also as air vents, these fittings ensure a flow of steam when initially warming through and raising steam in the boiler.

Sootblowers. Operated by steam or compressed air, they act to blow away soot and the products of combustion from the tube surfaces. Several are fitted in strategic places. The sootblower lance is inserted, soot is blown and the lance is withdrawn.

Water level gauges

The water level gauge provides a visible indication of the water level in the boiler in the region of the correct working level. If the water level were too high then water might pass out of the boiler and do serious damage to any equipment designed to accept steam. If the water level were too low then the heat transfer surfaces might become exposed to excessive temperatures and fail. Constant attention to the boiler water level is therefore essential. Due to the motion of the ship it is necessary to have a water level gauge at each end of the boiler to correctly observe the level.

Depending upon the boiler operating pressure, one of two basically different types of water level gauge will be fitted.

For boiler pressures up to a maximum of 17 bar a round glass tube type of water level gauge is used. The glass tube is connected to the boiler shell by cocks and pipes, as shown in Figure 4.11. Packing rings are positioned at the tube ends to give a tight seal and prevent leaks. A guard is usually placed around the tube to protect it from accidental damage and to avoid injury to any personnel in the vicinity if the tube shatters. The water level gauge is usually connected directly to the boiler. Isolating cocks are fitted in the steam and water passages, and a drain cock is also present. A ball valve is fitted below the tube to shut off the water should the tube break and water attempt to rush out.

For boiler pressures above 17 bar a plate-glass-type water level gauge is used. The glass tube is replaced by an assembly made up of glass plates within a metal housing, as shown in Figure 4.12. The assembly is made

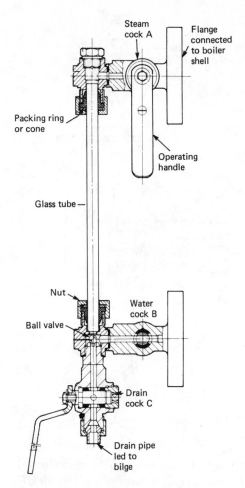

Figure 4.11 Tubular gauge glass

up as a 'sandwich' of front and back metal plates with the glass plates and a centre metal plate between. Joints are placed between the glass and the metal plate and a mica sheet placed over the glass surface facing the water and steam. The mica sheet is an effective insulation to prevent the glass breaking at the very high temperature. When bolting up this assembly, care must be taken to ensure even all-round tightening of the bolts. Failure to do this will result in a leaking assembly and possibly shattered glass plates.

In addition to the direct-reading water level gauges, remote-reading level indicators are usually led to machinery control rooms.

It is possible for the small water or steam passages to block with scale or dirt and the gauge will give an incorrect reading. To check that

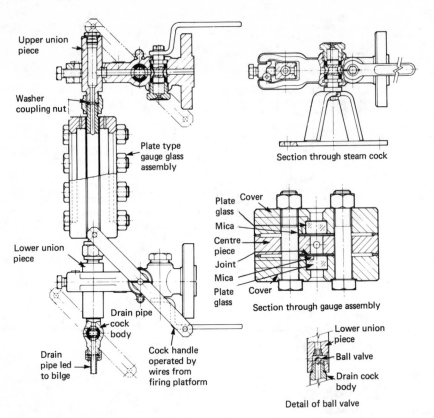

Upper union piece

Washer coupling nut

Plate type gauge glass assembly

Lower union piece

Drain pipe cock body

Drain pipe led to bilge

Cock handle operated by wires from firing platform

Section through steam cock

Plate glass
Cover
Mica
Centre piece
Joint
Mica
Plate glass
Cover

Section through gauge assembly

Lower union piece
Ball valve
Drain cock body

Detail of ball valve

Figure 4.12 Plate-type gauge glass

passages are clear a 'blowing through' procedure should be followed. Referring to Figure 4.11, close the water cock B and open drian cock C. The boiler pressure should produce a strong jet of steam from the drain. Cock A is now closed and Cock B opened. A jet of water should now pass through the drain. The absence of a flow through the drain will indicate that the passage to the open cock is blocked.

Safety valves

Safety valves are fitted in pairs, usually on a single valve chest. Each valve must be able to release all the steam the boiler can produce without the pressure rising by more than 10% over a set period.

Spring-loaded valves are always fitted on board ship because of their positive action at any inclination. They are positioned on the boiler drum in the steam space. The ordinary spring loaded safety valve is shown in Figure 4.13. The valve is held closed by the helical spring

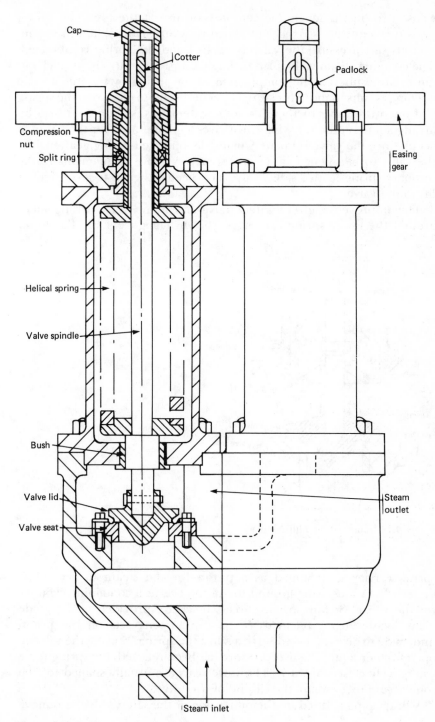

Figure 4.13 Ordinary spring-loaded safety valve

whose pressure is set by the compression nut at the top. The spring pressure, once set, is fixed and sealed by a Surveyor. When the steam exceeds this pressure the valve is opened and the spring compressed. The escaping steam is then led through a waste pipe up the funnel and out to atmosphere. The compression of the spring by the initial valve opening results in more pressure being necessary to compress the spring and open the valve further. To some extent this is countered by a lip arrangement on the valve lid which gives a greater area for the steam to act on once the valve is open. A manually operated easing gear enables the valve to be opened in an emergency. Various refinements to the ordinary spring-loaded safety valve have been designed to give a higher lift to the valve.

The improved high-lift safety valve has a modified arrangement around the lower spring carrier, as shown in Figure 4.14. The lower

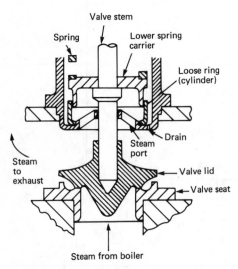

Figure 4.14 Improved high-lift safety valve

spring carrier is arranged as a piston for the steam to act on its underside. A loose ring around the piston acts as a containing cylinder for the steam. Steam ports or access holes are provided in the guide plate. Waste steam released as the valve opens acts on the piston underside to give increased force against the spring, causing the valve to open further. Once the overpressure has been relieved, the spring force will quickly close the valve. The valve seats are usually shaped to trap some steam to 'cushion' the closing of the valve.

A drain pipe is fitted on the outlet side of the safety valve to remove any condensed steam which might otherwise collect above the valve and stop it opening at the correct pressure.

Combustion

Combustion is the burning of fuel in air in order to release heat energy. For complete and efficient combustion the correct quantities of fuel and air must be supplied to the furnace and ignited. About 14 times as much air as fuel is required for complete combustion. The air and fuel must be intimately mixed and a small percentage of excess air is usually supplied to ensure that all the fuel is burnt. When the air supply is insufficient the fuel is not completely burnt and black exhaust gases will result.

Air supply

The flow of air through a boiler furnace is known as 'draught'. Marine boilers are arranged for forced draught, i.e. fans which force the air through the furnace. Several arrangements of forced draught are possible. The usual forced draught arrangement is a large fan which supplies air along ducting to the furnace front. The furnace front has an enclosed box arrangement, known as an 'air register', which can control the air supply. The air ducting normally passes through the boiler exhaust where some air heating can take place. The induced draught arrangement has a fan in the exhaust uptake which draws the air through the furnace. The balanced draught arrangement has matched forced draught and induced draught fans which results in atmospheric pressure in the furnace.

Fuel supply

Marine boilers currently burn residual low-grade fuels. This fuel is stored in double-bottom tanks from which it is drawn by a transfer pump up to settling tanks (Figure 4.15). Here any water in the fuel may settle out and be drained away.

The oil from the settling tank is filtered and pumped to a heater and then through a fine filter. Heating the oil reduces its viscosity and makes it easier to pump and filter. This heating must be carefully controlled otherwise 'cracking' or breakdown of the fuel may take place. A supply of diesel fuel may be available to the burners for initial firing or low-power operation of the boiler. From the fine filter the oil passes to the burner where it is 'atomised', i.e. broken into tiny droplets, as it enters the furnace. A recirculating line is provided to enable initial heating of the oil.

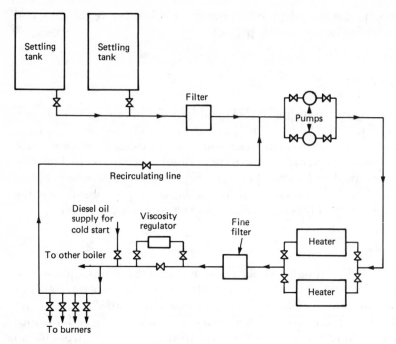

Figure 4.15 Boiler fuel-oil supply system

Fuel burning

The high-pressure fuel is supplied to a burner which it leaves as an atomised spray (Figure 4.16). The burner also rotates the fuel droplets by the use of a swirl plate. A rotating cone of tiny oil droplets thus leaves the burner and passes into the furnace. Various designs of burner exist, the one just described being known as a 'pressure jet burner' (Figure 4.16(a)). The 'rotating cup burner' (Figure 4.14(b)) atomises and swirls the fuel by throwing it off the edge of a rotating tapered cup. The 'steam blast jet burner', shown in Figure 4.16(c), atomises and swirls the fuel by spraying it into a high-velocity jet of steam. The steam is supplied down a central inner barrel in the burner.

The air register is a collection of flaps, vanes, etc., which surrounds each burner and is fitted between the boiler casings. The register provides an entry section through which air is admitted from the windbox. Air shut-off is achieved by means of a sliding sleeve or check. Air flows through parallel to the burner, and a swirler provides it with a rotating motion. The air is swirled in an opposite direction to the fuel to ensure adequate mixing (Figure 4.17(a)). High-pressure, high-output marine watertube boilers are roof fired (Figure 4.17(b)). This enables a long flame path and even heat transfer throughout the furnace.

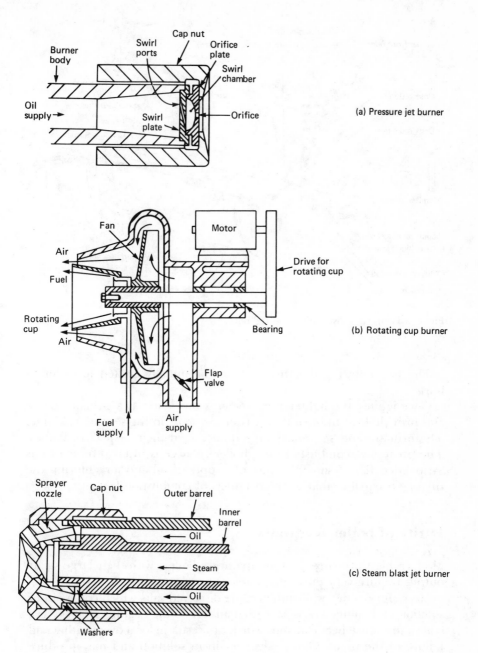

Figure 4.16 Types of burner

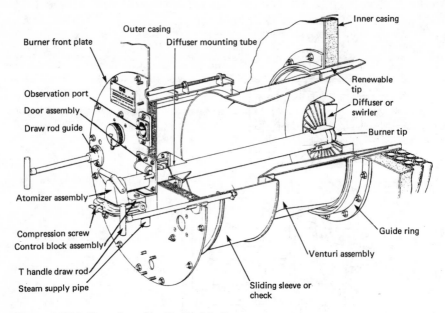

Figure 4.17(a) Air register for side-fired boiler

The fuel entering the furnace must be initially ignited in order to burn.

Once ignited the lighter fuel elements burn first as a primary flame and provide heat to burn the heavier elements in the secondary flame. The primary and secondary air supplies feed their respective flames. The process of combustion in a boiler furnace is often referred to as 'suspended flame' since the rate of supply of oil and air entering the furnace is equal to that of the products of combustion leaving.

Purity of boiler feedwater

Water Alkalinity

Modern high-pressure, high-temperature boilers with their large steam output require very pure feedwater.

Most 'pure' water will contain some dissolved salts which come out of solution on boiling. These salts then adhere to the heating surfaces as a scale and reduce heat transfer, which can result in local overheating and failure of the tubes. Other salts remain in solution and may produce acids which will attack the metal of the boiler. An excess of alkaline salts in a boiler, together with the effects of operating stresses, will produce a condition known as 'caustic cracking'. This is actual cracking of the metal which may lead to serious failure.

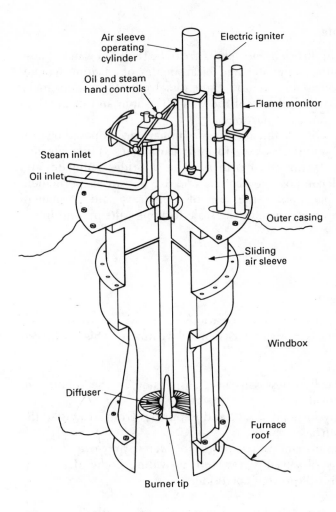

Figure 4.17(b) Air register for roof-fired boiler

The presence of dissolved oxygen and carbon dioxide in boiler feedwater can cause considerable corrosion of the boiler and feed systems. When boiler water is contaminated by suspended matter, an excess of salts or oil then 'foaming' may occur. This is a foam or froth which collects on the water surface in the boiler drum. Foaming leads to 'priming' which is the carry-over of water with the steam leaving the boiler drum. Any water present in the steam entering a turbine will do considerable damage.

Common impurities   c(ii) (iii) 1(e)

Various amounts of different metal salts are to be found in water. These include the chlorides, sulphates and bicarbonates of calcium, magnesium and, to some extent, sulphur. These dissolved salts in water make up what is called the 'hardness' of the water. Calcium and magnesium salts are the main causes of hardness.

The bicarbonates of calcium and magnesium are decomposed by heat and come out of solution as scale-forming carbonates. These alkaline salts are known as 'temporary hardness'. The chlorides, sulphates and nitrates are not decomposed by boiling and are known as 'permanent hardness'. Total hardness is the sum of temporary and permanent hardness and gives a measure of the scale-forming salts present in the boiler feedwater.

Feedwater treatment

Feedwater treatment deals with the various scale and corrosion causing salts and entrained gases by suitable chemical treatment. This is achieved as follows:

1. By keeping the hardness salts in a suspension in the solution to prevent scale formation.
2. By stopping any suspended salts and impurities from sticking to the heat transfer surfaces.
3. By providing anti-foam protection to stop water carry-over.
4. By eliminating dissolved gases and providing some degree of alkalinity which will prevent corrosion.

The actual treatment involves adding various chemicals into the feedwater system and then testing samples of boiler water with a test kit. The test kit is usually supplied by the treatment chemical manufacturer with simple instructions for its use.

For auxiliary boilers the chemicals added might be lime (calcium hydroxide) and soda (sodium carbonate). Alternatively caustic soda (sodium hydroxide) may be used on its own.

For high-pressure watertube boilers various phosphate salts are used, such as trisodium phosphate, disodium phosphate and sodium metaphosphate. Coagulants are also used which combine the scale-forming salts into a sludge and stop it sticking to the boiler surfaces. Sodium aluminate, starch and tannin are used as coagulants. Final de-aeration of the boiler water is achieved by chemicals, such as hydrazine, which combine with any oxygen present.

Boiler operation

The procedure adopted for raising steam will vary from boiler to boiler and the manufacturers' instructions should always be followed. A number of aspects are common to all boilers and a general procedure might be as follows.

Preparations

The uptakes should be checked to ensure a clear path for the exhaust gases through the boiler; any dampers should be operated and then correctly positioned. All vents, alarm, water and pressure gauge connections should be opened. The superheater circulating valves or drains should be opened to ensure a flow of steam through the superheater. All the other boiler drains and blow-down valves should be checked to ensure that they are closed. The boiler should then be filled to slightly below the working level with hot de-aerated water. The various header vents should be closed as water is seen to flow from them. The economiser should be checked to ensure that it is full of water and all air vented off.

The operation of the forced draught fan should be checked and where exhaust gas air heaters are fitted they should be bypassed. The fuel oil system should be checked for the correct positioning of valves, etc. The fuel oil should then be circulated and heated.

Raising steam

The forced draught fan should be started and air passed through the furnace for several minutes to 'purge' it of any exhaust gas or oil vapours. The air slides (checks) at every register, except the 'lighting up' burner, should then be closed. The operating burner can now be lit and adjusted to provide a low firing rate with good combustion. The fuel oil pressure and forced draught pressure should be matched to ensure good combustion with a full steady flame.

The superheater header vents may be closed once steam issues from them. When a drum pressure of about 210 kPa (2.1 bar) has been reached the drum air vent may be closed. The boiler must be brought slowly up to working pressure in order to ensure gradual expansion and to avoid overheating the superheater elements and damaging any refractory material. Boiler manufacturers usually provide a steam-raising diagram in the form of a graph of drum pressure against hours after flashing up.

The main and auxiliary steam lines should now be warmed through and then the drains closed. In addition the water level gauges should be blown through and checked for correct reading. When the steam pressure is about 300 kPa (3 bar) below the normal operating value the safety valves should be lifted and released using the easing gear.

Once at operating pressure the boiler may be put on load and the superheater circulating valves closed. All other vents, drains and bypasses should then be closed. The water level in the boiler should be carefully checked and the automatic water regulating arrangements observed for correct operation.

Chapter 5
Feed systems

The feed system completes the cycle between boiler and turbine to enable the exhausted steam to return to the boiler as feedwater. The feed system is made up of four basic items: the boiler, the turbine, the condenser and the feed pump. The boiler produces steam which is supplied to the turbine and finally exhausted as low-energy steam to the condenser. The condenser condenses the steam to water (condensate) which is then pumped into the boiler by the feed pump.

Other items are incorporated into all practical feed systems, such as a drain tank to collect the condensate from the condenser and provide a suction head for the feed pump. A make-up feed tank will provide additional feedwater to supplement losses or store surplus feed from the drain tank. In a system associated with an auxiliary boiler, as on a motor ship, the drain tank or hotwell will be open to the atmosphere. Such a feed system is therefore referred to as 'open feed'. In high-pressure watertube boiler installations no part of the feed system is open to the atmosphere and it is known as 'closed feed'.

Open feed system

An open feed system for an auxiliary boiler is shown in Figure 5.1. The exhaust steam from the various services is condensed in the condenser. The condenser is circulated by sea water and may operate at atmospheric pressure or under a small amount of vacuum. The condensate then drains under the action of gravity to the hotwell and feed filter tank. Where the condenser is under an amount of vacuum, extraction pumps will be used to transfer the condensate to the hotwell. The hotwell will also receive drains from possibly contaminated systems, e.g. fuel oil heating system, oil tank heating, etc. These may arrive from a drains cooler or from an observation tank. An observation tank, where fitted, permits inspection of the drains and their discharge to the oily bilge if contaminated. The feed filter and hotwell tank is arranged with internal baffles to bring about preliminary oil separation from any

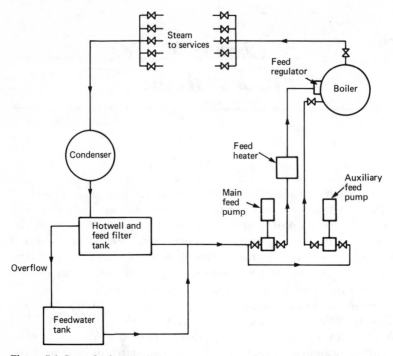

Figure 5.1 Open feed system

contaminated feed or drains. The feedwater is then passed through charcoal or cloth filters to complete the cleaning process. Any overflow from the hotwell passes to the feedwater tank which provides additional feedwater to the system when required. The hotwell provides feedwater to the main and auxiliary feed pump suctions. A feed heater may be fitted into the main feed line. This heater may be of the surface type, providing only heating, or may be of the direct contact type which will de-aerate in addition. De-aeration is the removal of oxygen in feedwater which can cause corrosion problems in the boiler. A feed regulator will control the feedwater input to the boiler and maintain the correct water level in the drum.

The system described above can only be said to be typical and numerous variations will no doubt be found, depending upon particular plant requirements.

Closed feed system

A closed feed system for a high pressure watertube boiler supplying a main propulsion steam turbine is shown in Figure 5.2.

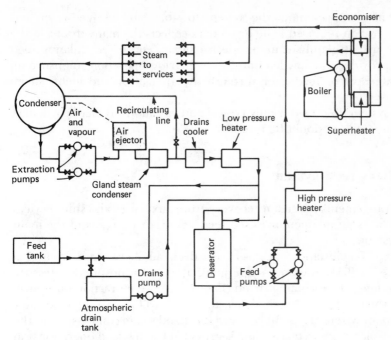

Figure 5.2 Closed feed system

The steam turbine will exhaust into the condenser which will be at a high vacuum. A regenerative type of condenser will be used which allows condensing of the steam with the minimum drop in temperature. The condensate is removed by an extraction pump and circulates through an air ejector.

The condensate is heated in passing through the air ejector. The ejector removes air from the condenser using steam-operated ejectors. The condensate is now circulated through a gland steam condenser where it is further heated. In this heat exchanger the turbine gland steam is condensed and drains to the atmospheric drain tank. The condensate is now passed through a low-pressure heater which is supplied with bled steam from the turbine. All these various heat exchangers improve the plant efficiency by recovering heat, and the increased feedwater temperature assists in the de-aeration process.

The de-aerator is a direct contact feed heater, i.e. the feedwater and the heating steam actually mix. In addition to heating, any dissolved gases, particularly oxygen, are released from the feedwater. The lower part of the de-aerator is a storage tank which supplies feedwater to the main feed pumps, one of which will supply the boiler's requirements.

The feedwater passes to a high-pressure feed heater and then to the economiser and the boiler water drum. An atmospheric drain tank and a feed tank are present in the system to store surplus feedwater and supply it when required. The drain tank collects the many drains in the system such as gland steam, air ejector steam, etc. A recirculating feed line is provided for low load and manœuvring operation to ensure an adequate flow of feedwater through the air ejector and gland steam condenser.

The system described is only typical and variations to meet particular conditions will no doubt be found.

Auxiliary feed system

The arrangements for steam recovery from auxiliaries and ship services may form separate open or closed feed sysems or be a part of the main feed system.

Where, for instance, steam-driven deck auxiliaries are in use, a separate auxiliary condenser operating at about atmospheric pressure will condense the incoming steam (Figure 5.3). An extraction pump will supply the condensate to an air ejector which will return the feedwater to the main system at a point between the gland steam condenser and the drains cooler. A recirculating line is provided for low-load operation and a level controller will maintain a condensate level in the condenser.

Where contamination of the feedwater may be a problem, a separate feed system for a steam-to-steam generator can be used (Figure 5.4). Low-pressure steam from the generator is supplied to the various services, such as fuel oil heating, and the drains are returned to the hotwell. Feed pumps supply the feed to a feed heater, which also acts as

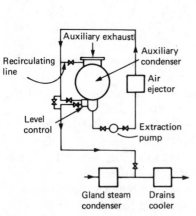

Figure 5.3 Auxiliary feed system

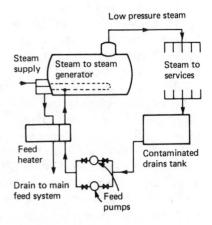

Figure 5.4 Steam-to-steam generator feed system

a drains cooler for the heating steam supplied to the generator. From the feed heater, the feedwater passes into the steam-to-steam generator.

Packaged feed systems are also available from a number of manufacturers. With this arrangement the various system items are mounted on a common base or bedplate. The complete feed system may be packaged or a number of the items.

System components

Condenser

The condenser is a heat exchanger which removes the latent heat from exhaust steam so that it condenses and can be pumped back into the boiler. This condensing should be achieved with the minimum of under-cooling, i.e. reduction of condensate temperature below the steam temperature. A condenser is also arranged so that gases and vapours from the condensing steam are removed.

An auxiliary condenser is shown in Figure 5.5. The circular cross-section shell is provided with end covers which are arranged for a two-pass flow of sea water. Sacrificial corrosion plates are provided in the water boxes. The steam enters centrally at the top and divides into two paths passing through ports in the casing below the steam inlet hood. Sea water passing through the banks of tubes provides the cooling surface for condensing the steam. The central diaphragm plate supports the tubes and a number of stay rods in turn support the diaphragm plate. The condensate is collected in a sump tank below the tube banks. An air suction is provided on the condenser shell for the withdrawal of gases and vapours released by the condensing steam.

Main condensers associated with steam turbine propulsion machinery are of the regenerative type. In this arrangement some of the steam bypasses the tubes and enters the condensate sump as steam. The condensate is thus reheated to the same temperature as the steam, which increases the efficiency of the condenser. One design of regenerative condenser is shown in Figure 5.6. A central passage enables some of the steam to pass to the sump, where it condenses and heats the condensate. A baffle plate is arranged to direct the gases and vapours towards the air ejector. The many tubes are fitted between the tube plates at each end and tube support plates are arranged between. The tubes are circulated in two passes by sea water.

Extraction pump

The extraction pump is used to draw water from a condenser which is under vacuum. The pump also provides the pressure to deliver the feed water to the de-aerator or feed pump inlet.

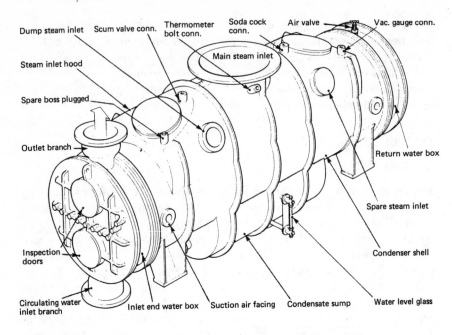

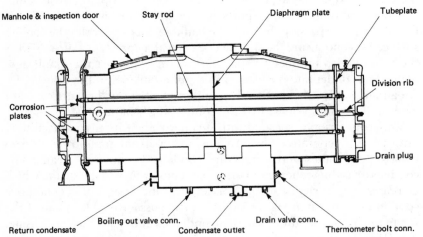

Figure 5.5 Auxiliary condenser

Extraction pumps are usually of the vertical shaft, two stage, centrifugal type, as described in Chapter 6. These pumps require a specified minimum suction head to operate and, usually, some condensate level control system in the condenser. The first-stage impeller receives water which is almost boiling at the high vacuum conditions present in the suction pipe. The water is then discharged at a

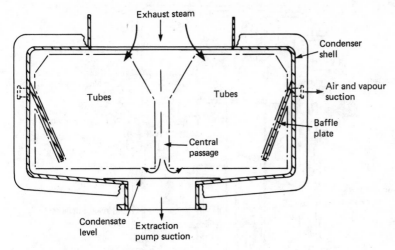

Figure 5.6 Regenerative condenser

slight positive pressure to the second-stage impeller which provides the necessary system pressure at outlet.

Where the condenser sump level is allowed to vary or maintained almost dry, a self-regulating extraction pump must be used. This regulation takes the form of cavitation which occurs when the suction head falls to a very low value. Cavitation is the formation and collapse of vapour bubbles which results in a fall in the pump discharge rate to zero. As the suction head improves the cavitation gradually ceases and the pump begins to discharge again. Cavitation is usually associated with damage (see Chapter 11 with reference to propellers) but at the low-pressure conditions in the pump no damage occurs. Also, the impeller may be designed so that the bubble collapse occurs away from the impeller, i.e. super-cavitating.

Air ejector

The air ejector draws out the air and vapours which are released from the condensing steam in the condenser. If the air were not removed from the system it could cause corrosion problems in the boiler. Also, air present in the condenser would affect the condensing process and cause a back pressure in the condenser. The back pressure would increase the exhaust steam pressure and reduce the thermal efficiency of the plant.

A two-stage twin-element air ejector is shown in Figure 5.7. In the first stage a steam-operated air ejector acts as a pump to draw in the air and vapours from the condenser. The mixture then passes into a condensing unit which is circulated by feedwater. The feedwater is heated and the

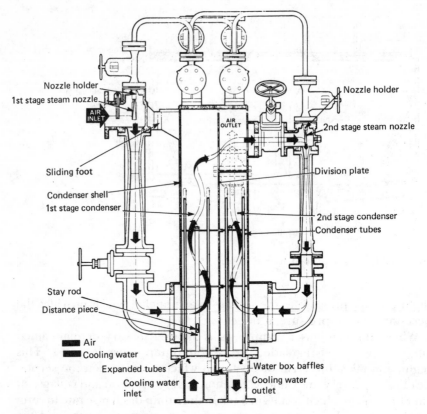

Figure 5.7 Air ejector

steam and gases are mostly condensed. The condensed vapours and steam are returned to the main condenser via a drain and the remaining air and gases pass to the second stage where the process is repeated. Any remaining air and gases are released to the atmosphere via a vacuum-retaining valve. The feed water is circulated through U-tubes in each of the two stages. A pair of ejectors are fitted to each stage, although only one of each is required for satisfactory operation of the unit.

Heat exchangers

The gland steam condenser, drains cooler and low-pressure feed heater are all heat exchangers of the shell and tube type. Each is used in some particular way to recover heat from exhaust steam by heating the feedwater which is circulated through the units.

The gland steam condenser collects steam, vapour and air from the turbine gland steam system. These returns are cooled by the circulating feedwater and the steam is condensed. The condensate is returned to the system via a loop seal or some form of steam trap and any air present is discharged into the atmosphere. The feedwater passes through U-tubes within the shell of the unit.

The drains cooler receives the exhaust drains from various auxiliary services and condenses them: the condensate is returned to the feed system. The circulating feedwater passes through straight tubes arranged in tube plates in the drains cooler. Baffles or diaphragm plates are fitted to support the tubes and also direct the flow of the exhaust drains over the outside surface of the tubes (Figure 5.8).

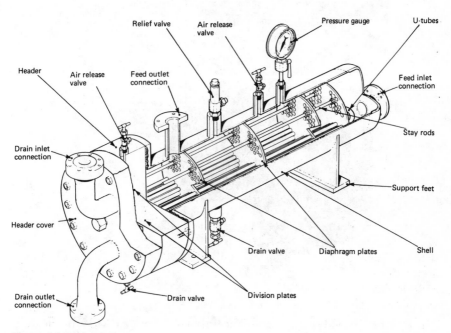

Figure 5.8 Drains cooler

The low-pressure feed heater is supplied with steam usually bled from the low-pressure turbine casing. The circulating feed is heated to assist in the de-aeration process. The bleeding-off of steam from the turbine improves plant thermal efficiency as well as reducing turbine blade heights in the final rows because of the reduced mass of steam flowing. Either straight or U-tube construction may be used with single or multiple passes of feedwater.

De-aerator

The de-aerator completes the air and vapour removal process begun in the condenser. It also functions as a feed heater, but in this case operates by direct contact. The feedwater is heated almost to the point of boiling, which releases all the dissolved gases which can then be vented off.

One type of de-aerator is shown in Figure 5.9. The incoming feedwater passes through a number of spray valves or nozzles: the water

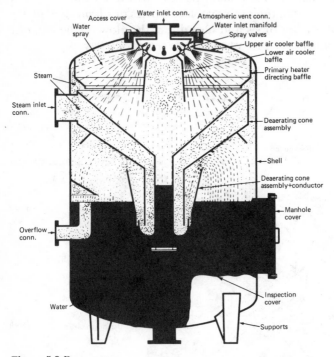

Figure 5.9 De-aerator

spray thus provides a large surface area for contact with the heating system. Most of the feedwater will then fall onto the upper surface of the de-aerating cone where it is further heated by the incoming steam. The feedwater then enters the central passage and leaves through a narrow opening which acts as an eductor or ejector to draw steam through with the feed. The feedwater and condensed steam collect in the storage tank which forms the base of the de-aerator. The heating steam enters the de-aerator and circulates throughout, heating the feedwater and being condensed in its turn to combine with the feedwater. The released gases leave through a vent connection and pass to a vent condenser or devaporiser. Any water vapour will be condensed and returned. The devaporiser is circulated by the feedwater before it enters the de-aerator.

The de-aerator feedwater is very close to the steam temperature at the same pressure and will, if subjected to any pressure drop, 'flash-off' into steam. This can result in 'gassing', i.e. vapour forming in the feed pump suction. To avoid this problem, the de-aerator is mounted high up in the machinery space to give a positive suction head to the feed pumps. Alternatively a booster or extraction pump may be fitted at the de-aerator outlet.

Feed pump

The feed pump raises the feedwater to a pressure high enough for it to enter the boiler.

For auxiliary boilers, where small amounts of feedwater are pumped, a steam-driven reciprocating positive displacement pump may be used. This type of pump is described in Chapter 6. Another type of feed pump often used on package boiler installations is known as an 'electro

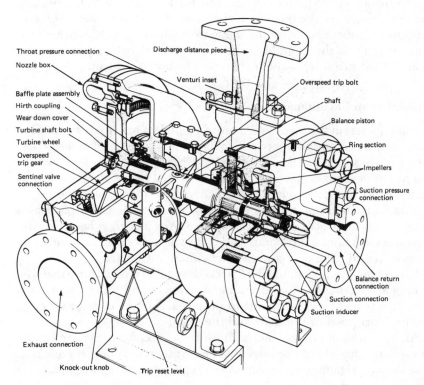

Figure 5.10 Turbo-feed pump

feeder'. This is a multi-stage centrifugal pump driven by a constant-speed electric motor. The number of stages is determined by the feed quantity and discharge pressure.

Steam turbine-driven feed pumps are usual with high-pressure watertube boiler installations. A typical turbo-feed pump is shown in Figure 5.10. The two-stage horizontal centrifugal pump is driven by an impulse turbine, the complete assembly being fitted into a common casing. The turbine is supplied with steam directly from the boiler and exhausts into a back-pressure line which can be used for feed heating. The pump bearings are lubricated by filtered water which is tapped off from the first-stage impeller. The feed discharge pressure is maintained by a governor, and overspeed protection trips are also provided.

High-pressure feed heater

The high-pressure feed heater is a heat exchanger of the shell and tube type which further heats the feedwater before entry to the boiler. Further heat may be added to the feedwater without its becoming steam since its pressure has now been raised by the feed pump.

The incoming feedwater circulates through U-tubes with the heating steam passing over the outside of the tubes. Diaphragm plates serve to support the tubes and direct the steam through the heater. A steam trap ensures that all the heating steam is condensed before it leaves the heater. Bled steam from the turbine will be used for heating.

Operation and maintenance

During operation the feed system must maintain a balance between feed input and steam output, together with a normal water level in the boiler. The control system used is described in Chapter 15.

The condenser sea water boxes are protected by sacrificial mild steel plates which must be renewed regularly. The tube plates should be examined at the same time to ensure no erosion has taken place as a result of too high a circulating water speed. Any leaking tubes will cause feedwater contamination, and where this is suspected the condenser must be tested. The procedure is mentioned in Chapter 7.

Extraction pumps should be checked regularly to ensure that the sealing arrangements are preventing air from entering the system. It is usual with most types of glands to permit a slight leakage of water to ensure lubrication of the shaft and the gland.

Air ejectors will operate inefficiently if the ejector nozzles are coated or eroded. They should be inspected and cleaned or replaced regularly. The vacuum retaining valve should be checked for air tightness and also the ejector casing.

The various heat exchangers should be checked regularly for tube leakages and also the cleanliness of the heat-exchange surfaces.

The operation of the reciprocating positive displacement pump is described in Chapter 6. Turbo-feed pumps are started with the discharge valve closed in order to build up pressure rapidly and bring the hydraulic balance into operation. The turbine driving the pump will require warming through with the drains open before running up to speed and then closing the drains. The turbine overspeed trip should be checked regularly for correct operation and axial clearances should be measured, usually with a special gauge.

Chapter 6
Pumps and pumping systems

At any one time in a ship's machinery space there will be a considerable variety of liquids on the move. The lengths of pipework will cover many kilometres, the systems are often interconnecting and most pumps are in pairs. The engineer must be familiar with each system from one end to the other, knowing the location and use of every single valve. The various systems perform functions such as cooling, heating, cleaning and lubricating of the various items of machinery. Each system can be considered comprised of _pumps, piping, valves_ and _fittings_, which will now be examined in turn.

Pumps

A pump is a machine used to raise liquids from a low point to a high point. Alternatively it may simply provide the liquid with an increase in energy enabling it to flow or build up a pressure. The pumping action can be achieved in various ways according to the type of pump employed. The arrangement of pipework, the liquid to be pumped and its purpose will result in certain system requirements or characteristics that must be met by the pump.

A pumping system on a ship will consist of suction piping, a pump and discharge piping (Figure 6.1). The system is arranged to provide a positive pressure or head at some point and discharge the liquid. The pump provides the energy to develop the head and overcome any losses in the system. Losses are mainly due to friction within the pipes and the difference between the initial and final liquid levels. The total system losses, H_{TOTAL} are found as follows:

$$H_{TOTAL} = H_{FRSUCT} + H_{FRDIS} + H_{DISTANK} + H_{SUCTTANK}$$

where H_{FRSUCT} = friction head loss in suction piping
H_{FRDIS} = friction head loss in discharge piping
$H_{DISTANK}$ = height of discharge tank level above pump

112

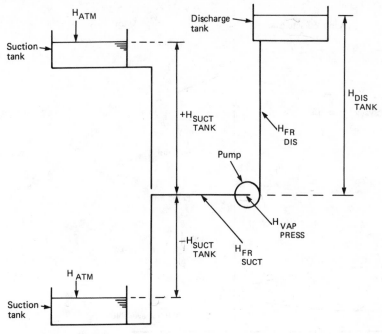

Figure 6.1 Basic pumping system

$$H_{SUCTTANK} = \text{ height of suction tank level above pump}$$
(negative when tank level is below pump suction)

All values are in metres of liquid.

The system head loss–flow characteristic can be drawn as shown in Figure 6.2. The system flow rate or capacity will be known and the pump manufacturer will provide a head–flow characteristic for his equipment which must be matched to the system curve. To obtain the best operating conditions for the pump it should operate over its range of maximum efficiency. A typical centrifugal pump characteristic is shown in Figure 6.2.

An important consideration, particularly when drawing liquids from below the pump, is the suction-side conditions of the system. The determination of Net Positive Suction Head (NPSH) is undertaken for both the system and the pump. Net Positive Suction Head is the difference between the absolute pump inlet pressure and the vapour pressure of the liquid, and is expressed in metres of liquid. Vapour pressure is temperature dependent and therefore NPSH should be given for the operating temperature of the liquid. The NPSH available in the system is found as follows:

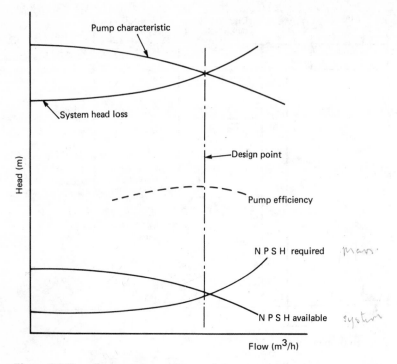

Figure 6.2 Overall pump system characteristics

$$\text{NPSH}_{\text{AVAIL}} = \underbrace{H_{\text{ATM}} + H_{\text{SUCTTANK}} - H_{\text{FRSUCT}}}_{\text{absolute pump inlet pressure}} - H_{\text{VAPPRESS}}$$

where H_{ATM} = atmospheric pressure
      H_{SUCTTANK} = tank level from pump (negative when tank level
                    is below pump)
      H_{FRSUCT} = friction head loss in suction piping
      H_{VAPPRESS} = liquid vapour pressure

The above values are usually expressed in metres head of sea water.
The pump manufacturer provides a NPSH required characteristic for
the pump which is also in metres head of sea water (Figure 6.2). The
pump and system must be matched in terms of NPSH such that NPSH
required is always greater than NPSH available. An insufficient value of
NPSH required will result in cavitation, i.e. the forming and collapsing
of bubbles in the liquid, which will affect the pumping operation and
may damage the pump.

Pump types

There are three main classes of pump in marine use: *displacement, axial flow* and *centrifugal*. A number of different arrangements are possible for displacement and centrifugal pumps to meet particular system characteristics.

Displacement

The displacement pumping action is achieved by the reduction or increase in volume of a space causing the liquid (or gas) to be physically moved. The method employed is either a piston in a cylinder using a reciprocating motion, or a rotating unit using vanes, gears or screws.

A reciprocating displacement pump is shown diagrammatically in Figure 6.3, to demonstrate the operating principle. The pump is

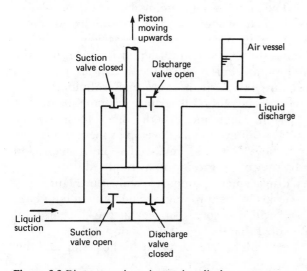

Figure 6.3 Diagrammatic reciprocating displacement pump

double-acting, that is liquid is admitted to either side of the piston where it is alternately drawn in and discharged. As the piston moves upwards, suction takes place below the piston and liquid is drawn in, the valve arrangement ensuring that the discharge valve cannot open on the suction stroke. Above the piston, liquid is discharged and the suction valve remains closed. As the piston travels down, the operations of suction and discharge occur now on opposite sides.

An air vessel is usually fitted in the discharge pipework to dampen out the pressure variations during discharge. As the discharge pressure rises the air is compressed in the vessel, and as the pressure falls the air expands. The peak pressure energy is thus 'stored' in the air and returned to the system when the pressure falls. Air vessels are not fitted on reciprocating boiler feed pumps since they may introduce air into the de-aerated feedwater.

A relief valve is always fitted between the pump suction and discharge chambers to protect the pump should it be operated with a valve closed in the discharge line.

Reciprocating displacement pumps are self priming, will accept high suction lifts, produce the discharge pressure required by the system and can handle large amounts of vapour or entrained gases. They are, however, complicated in construction with a number of moving parts requiring attention and maintenance.

When starting the pump the suction and discharge valves must be opened. It is important that no valves in the discharge line are closed, otherwise either the relief valve will lift or damage may occur to the pump when it is started. The pump is self priming, but where possible to reduce wear or the risk of seizure it should be flooded with liquid before starting. An electrically driven pump needs only to be switched on, when it will run erratically for a short period until liquid is drawn into the pump. A steam driven pump will require the usual draining and warming-through procedure before steam is gradually admitted.

Most of the moving parts in the pump will require examination during overhaul. The pump piston, rings and cylinder liner must also be thoroughly checked. Ridges will eventually develop at the limits of the piston ring travel and these must be removed. The suction and discharge valves must be refaced or ground in as required.

Two different rotary displacement pumps are shown in Figure 6.4. The action in each case results in the trapping of a quantity of liquid (or air) in a volume or space which becomes smaller at the discharge or outlet side. It should be noted that the liquid does not pass between the screw or gear teeth as they mesh but travels between the casing and the teeth.

The starting procedure is similar to that for the reciprocating displacement pump. Again a relief valve will be fitted between suction and discharge chambers. The particular maintenance problem with this type of pump is the shaft sealing where the gland and packing arrangement must be appropriate for the material pumped. The rotating vane type will suffer wear at a rate depending upon the liquid pumped and its freedom from abrasive or corrosive substances. The screw pump must be correctly timed and if stripped for inspection care should be taken to assemble the screws correctly.

A special type of rotary displacement pump has a particular application in steering gear and is described in Chapter 12.

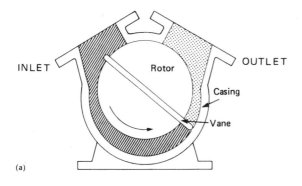

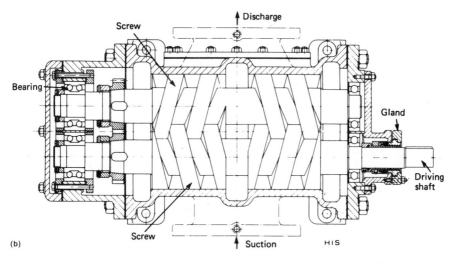

Figure 6.4 Rotary displacement pumps: (a) rotary vane displacement pump, (b) screw displacement pump

Axial-flow pump

An axial-flow pump uses a screw propeller to axially accelerate the liquid. The outlet passages and guide vanes are arranged to convert the velocity increase of the liquid into a pressure.

A reversible axial flow pump is shown in Figure 6.5. The pump casing is split either horizontally or vertically to provide access to the propeller. A mechanical seal prevents leakage where the shaft leaves the casing. A thrust bearing of the tilting pad type is fitted on the drive shaft. The prime mover may be an electric motor or a steam turbine.

The axial flow pump is used where large quantities of water at a low head are required, for example in condenser circulating. The efficiency

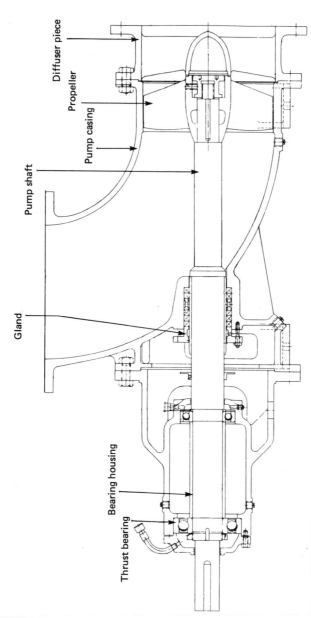

Figure 6.5 Axial-flow pump

is equivalent to a low lift centrifugal pump and the higher speeds possible enable a smaller driving motor to be used. The axial-flow pump is also suitable for supplementary use in a condenser scoop circulating system since the pump will offer little resistance to flow when idling. With scoop circulation the normal movement of the ship will draw in water; the pump would be in use only when the ship was moving slowly or stopped.

Centrifugal pump

In a centrifugal pump liquid enters the centre or eye of the impeller and flows radially out between the vanes, its velocity being increased by the impeller rotation. A diffuser or volute is then used to convert most of the kinetic energy in the liquid into pressure. The arrangement is shown diagrammatically in Figure 6.6.

A vertical, single stage, single entry, centrifugal pump for general marine duties is shown in Figure 6.7. The main frame and casing, together with a motor support bracket, house the pumping element assembly. The pumping element is made up of a top cover, a pump shaft, an impeller, a bearing bush and a sealing arrangement around the shaft. The sealing arrangement may be a packed gland or a mechanical seal and the bearing lubrication system will vary according to the type of seal. Replaceable wear rings are fitted to the impeller and the casing. The motor support bracket has two large apertures to provide access to the pumping element, and a coupling spacer is fitted between the motor and pump shaft to enable the removal of the pumping element without disturbing the motor.

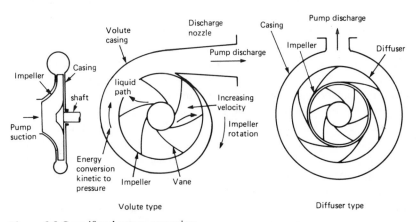

Figure 6.6 Centrifugal pump operation

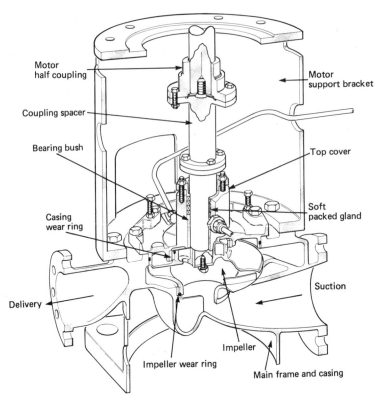

Figure 6.7 Single-entry centrifugal pump

Other configurations of centrifugal pumps are used for particular duties or to meet system requirements. A vertical single stage double-entry centrifugal pump is shown in Figure 6.8. The incoming liquid enters the double impeller from the top and the bottom and passes into the volute casing for discharge. A double-entry pump has a lower NPSH required characteristic which will have advantages in poor suction conditions. It should be noted that different impeller sizes can be fitted into a basic pumping element. This enables various discharge head characteristics to be provided for the same basic pump frame.

A vertical multi-stage single-entry centrifugal pump used for deep-well cargo pumping is shown in Figure 6.9. This can be considered as a series of centrifugal pumps arranged to supply one another in series and thus progressively increase the discharge pressure. The pump drive is located outside the tank and can be electric, hydraulic or any appropriate means suitable for the location.

A diffuser is fitted to high-pressure centrifugal pumps. This is a ring fixed to the casing, around the impeller, in which there are passages

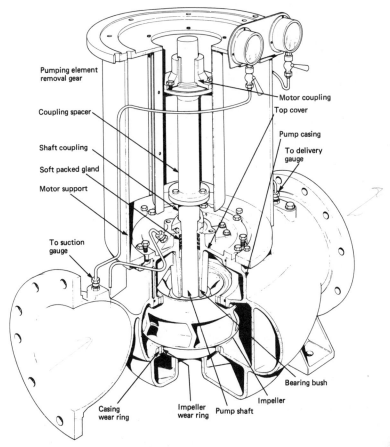

Pumping element
removal gear

Coupling spacer

Shaft coupling

Soft packed gland

Motor support

To suction
gauge

Motor coupling

Top cover

Pump casing

To delivery
gauge

Bearing bush

Casing
wear ring

Impeller
wear ring

Pump shaft

Impeller

Figure 6.8 Double-entry centrifugal pump

formed by vanes. The passages widen out in the direction of liquid flow
and act to convert the kinetic energy of the liquid into pressure energy.

Hydraulic balance arrangements are also usual. Some of the
high-pressure discharge liquid is directed against a drum or piston
arrangement to balance the discharge liquid pressure on the impeller
and thus maintain it in an equilibrium position.

Centrifugal pumps, while being suitable for most general marine
duties, are not self priming and require some means of removing air
from the suction pipeline and filling it with liquid. Where the liquid to be
pumped is at a level higher than the pump, opening an air cock near the
pump suction will enable the air to be forced out as the pipeline fills up
under the action of gravity. If the pump is below sea water level, and sea
water priming is permissible in the system, then opening a sea water
injection valve and the air cock on the pump will effect priming.

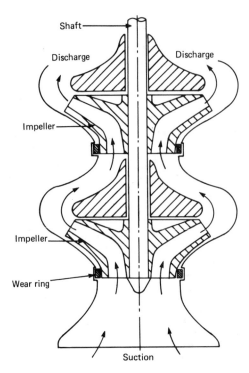

Shaft

Discharge   Discharge

Impeller

Impeller

Wear ring

Suction

Note only two stages are shown

Figure 6.9 Multi-stage centrifugal pump

Alternatively an air pumping unit can be provided to individual pumps or as a central priming system connected to several pumps.

The water ring or liquid ring primer can be arranged as an individual unit mounted on the pump and driven by it, or as a motor driven unit mounted separately and serving several pumps. The primer consists of an elliptical casing in which a vaned rotor revolves. The rotor may be separate from the hub and provide the air inlet and discharge ports as shown in Figure 6.10. Alternatively another design has the rotor and hub as one piece with ports on the cover. The rotor vanes revolve and force a ring of liquid to take up the elliptical shape of the casing. The water ring, being elliptical, advances and recedes from the central hub, causing a pumping action to occur. The suction piping system is connected to the air inlet ports and the suction line is thus primed by the removal of air. The air removed from the system is discharged to atmosphere. A reservoir of water is provided to replenish the water ring when necessary.

When starting a centrifugal pump the suction valve is opened and the discharge valve left shut: then the motor is started and the priming unit will prime the suction line. Once the pump is primed the delivery valve

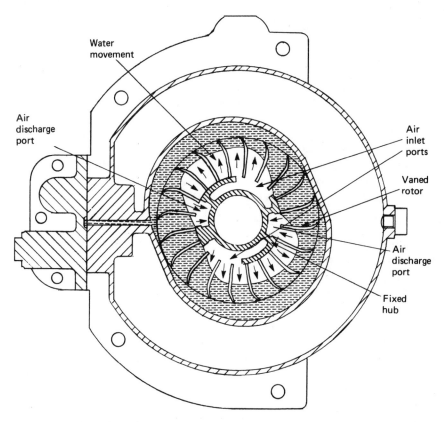

Figure 6.10 Water-ring primer

can be slowly opened and the quantity of liquid can be regulated by opening or closing the delivery valve. When stopping the pump the delivery valve is closed and the motor stopped.

Regular maintenance on the machine will involve attention to lubrication of the shaft bearing and ensuring that the shaft seal or gland is not leaking liquid. Unsatisfactory operation or loss of performance may require minor or major overhauls. Common faults, such as no discharge, may be a result of valves in the system being shut, suction strainers blocked or other faults occurring in the priming system. Air leaks in the suction piping, a choked impeller or too tight a shaft gland can all lead to poor performance.

When dismantling the pump to remove the pumping element any priming pipes or cooling water supply pipes must be disconnected. Modern pumps have a coupling spacer which can be removed to enable the pumping element to be withdrawn without disturbing the motor: the impeller and shaft can then be readily separated for examination. The shaft bearing bush together with the casing and impeller wear rings should be examined for wear.

Ejectors

An ejector is a form of pump which has no moving parts. An ejector arrangement is shown in Figure 5.7, Chapter 5. A high-pressure liquid or a vapour such as steam discharges from a nozzle as a high-velocity jet and entrains any gases or liquids surrounding the nozzle. The mixture enters a converging–diverging diffuser tube where some of the kinetic energy is converted into pressure energy. Ejectors can be arranged as single or multi-stage units and have particular applications, for example the air ejector in a closed feed system (see Chapter 5).

Piping systems

A ship's machinery space contains hundreds of metres of piping and fittings. The various systems are arranged to carry many different liquids at various temperatures and pressures. The influences of operational and safety requirements, as well as legislation, result in somewhat complicated arrangements of what are a few basic fittings. Valves, strainers, branch pipes, etc., are examples of fittings which are found in a pipe system.

Pipes

Machinery space pipework is made up of assorted straight lengths and bends joined by flanges with an appropriate gasket or joint between, or very small-bore piping may use compression couplings. The piping material will be chosen to suit the liquid carried and the system conditions. Some examples are given in Table 6.1.

Where piping is to be galvanised, the completed pipe with all joints fully welded is to be hot dipped galvanised. The pipes are supported

Table 6.1 Pipework material

System	Material
Waste steam	Carbon steel to BS 3601
SW circulating	Aluminium brass
Wash deck and firemain	Carbon steel to BS 3601 – galvanised
Bilge and ballast	Carbon steel to BS 3601 – galvanised
Control air	Copper
Starting air	Carbon steel to BS 3602

and held in by hangers or pipe clips in such a way as to minimise vibration. Steam pipes or pipes in systems with considerable temperature variation may be supported on spring hangers which permit a degree of movement. An alternative to spring hangers is the use of expansion loops of piping or an expansion joint.

Valves

Valves are provided in a piping system to regulate or stop the liquid flow. Various types exist with their associated particular function or advantages.

Cock

A cock is used in small-bore pipework and is joined to adjacent pipework by a compression coupling. A cock can restrict or close an internal passage by moving a central plug, usually by an external lever. An example of a straight-through cock is given in Figure 6.11.

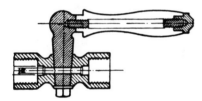

Figure 6.11 Cock

Globe valve

A globe valve has a somewhat spherical body enclosing the valve seat and valve disc (Figure 6.12). Flanges are provided at either side for connecting to adjacent pipework, and internal passages guide the liquid flow through the valve seat. Liquid flow is always arranged to come from below the valve seat so that the upper chamber is not pressurised when the valve is closed. A screw lift valve arrangement is shown where the spindle is joined to the valve disc. A gland with appropriate packing surrounds the spindle where it leaves the valve bonnet. The upper part of the spindle is threaded and passes through a similarly threaded bridge piece. A circular handwheel is used to turn the spindle and raise or lower the valve disc. The valve disc and seat are a perfect match and may be flat or, more commonly, mitred. The material for both is often provided with a very hard stellite coating. Globe valves also exist in a right-angled form where the inlet and exit flanges are at 90° to each other.

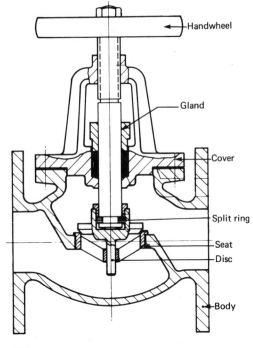

Figure 6.12 Globe valve

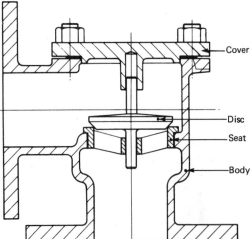

Figure 6.13 Non-return valve

Non-return or check valves are arranged in various pipelines to prevent reverse flow. Where the valve disc is not attached to the spindle it is known as screw-down non-return (SDNR). The valve disc in such a valve must have some form of guide or wings to ensure it can reseat correctly when screwed closed. Non-return valves are sometimes arranged without spindles, in which case they are liquid operated and can not be manually closed (Figure 6.13). A free lifting valve may be used or a hinged flap.

Gate valve

A gate valve should be fully open or closed; it is not suitable for flow control. When open it provides a clear full-bore internal passage for the liquid since the valve or gate is raised clear (Figure 6.14). The spindle is

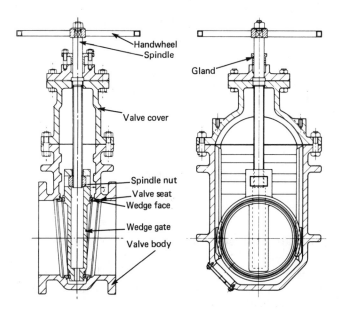

Figure 6.14 Gate valve

threaded over its lower portion and when turned causes the gate to raise or lower. The gate may be parallel or wedge-shaped in section fitting against a matching seat. Larger valves have replaceable seat rings and gate facings.

Relief valves

Excess pressure is avoided in pipe systems by the use of relief valves. The valve disc is held closed by a spring arrangement on the stem (Figure 6.15). The spring compression can be adjusted to enable the valve to open at the appropriate pressure. Boiler safety valves are a special case of relief valve and are described in Chapter 4.

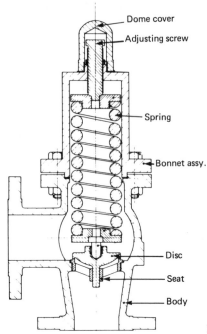

Dome cover

Adjusting screw

Spring

Bonnet assy.

Disc

Seat

Body

Figure 6.15 Relief valve

Quick-closing valves

Oil tank suction valves are arranged for rapid closing from a remote point by the use of quick-closing valves. The collapsing of the 'bridge' results in the valve closing quickly under the combined effects of gravity and an internal spring. A manually operated wire or a hydraulic cylinder can be used to collapse the bridge.

Valve chests

Valve chests are a series of valves all built into a single block or manifold. Various arrangements of suction and discharge connections are possible with this assembly. A particular application of this assembly is the change-over chest (Figure 6.16). Two interchangeable pieces are provided which enable a tank suction to pass to either the ballast main or the oil transfer main, but not both. Such an arrangement is essential where tanks can be used for either water ballast or oil.

Other fittings

Mud boxes are fitted into the machinery space bilge suction piping. The mud box is a coarse strainer with a straight tailpipe down to the bilge

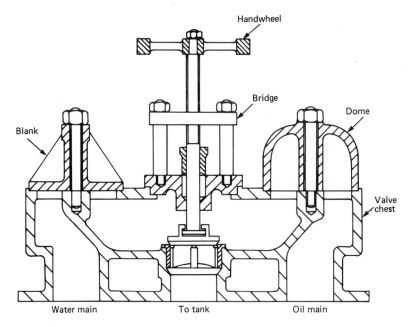

Figure 6.16 Change-over chest

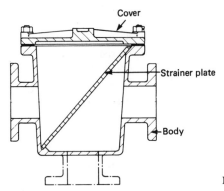

Figure 6.17 Mud box

(Figure 6.17). To enable the internal perforated plate to be cleaned when necessary, the lid of the mud box is easily removed without disconnecting any pipework.

Suction pipes in tanks should be arranged with a bell mouth or foot. The bell end or foot should provide an inlet area of about one-and-a-half times the pipe area. It should also be a sufficient distance from the bottom plating and nearby structure to provide a free suction area, again about one-and-a-half times the pipe area.

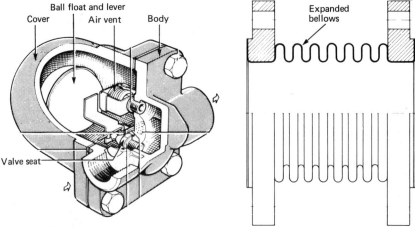

Figure 6.18 Steam trap

Figure 6.19 Expansion bellows piece

A steam trap does as its name implies and permits only the passage of condensed steam. It operates automatically and is situated in steam drain lines. Various designs are available utilising mechanical floats which, when floating in condensate, will enable the condensate to discharge (Figure 6.18). Other designs employ various types of thermostat to operate the valve which discharges the condensate.

An expansion piece is fitted in a pipeline which is subject to considerable temperature variations. One type consists of a bellows arrangement which will permit movement in several directions and absorb vibration (Figure 6.19). The fitting must be selected according to the variation in system temperatures and installed to permit the expansion and contraction required in the system.

Drains are provided in pipelines and usually have small cocks to open or close them. It is essential that certain pipelines are drained regularly, particularly in steam systems. When steam is admitted to a pipeline containing a reasonable surface of water it will condense and a partial vacuum occur: the water will then be drawn along the pipe until it meets a bend or a closed valve. The impact of the moving water in the pipework will create large forces known as 'water hammer', which can result in damage to pipework and fittings.

Bilge and ballast systems

The bilge system and the ballast system each have particular functions to perform, but are in many ways interconnected.

Bilge system

The bilge main is arranged to drain any watertight compartment other than ballast, oil or water tanks and to discharge the contents overboard. The number of pumps and their capacity depend upon the size, type and service of the vessel. All bilge suctions must be fitted with suitable strainers, which in the machinery space would be mud boxes positioned at floorplate level for easy access. A vertical drop pipe would lead down to the bilge.

The emergency bilge suction or bilge injection valve is used to prevent flooding of the ship. It is a direct suction from the machinery space bilge which is connected to the largest capacity pump or pumps. An emergency bilge pump is required for passenger ships but may also be fitted as an extra on cargo ships. It must be a completely independent unit capable of operating even if submerged. A centrifugal pump with a priming device is usually used, driven by an electric motor housed in an air bell. The power supply is arranged from the emergency generator.

A typical system is shown in Figure 6.20. The various pumps and lines are interconnected to some extent so that each pump can act as an alternative or standby for another.

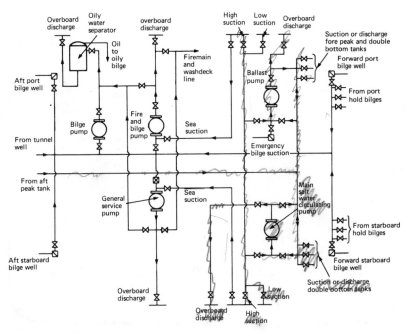

Figure 6.20 Bilge and ballast systems

Ballast systems

The ballast system is arranged to ensure that water can be drawn from any tank or the sea and discharged to any other tank or the sea as required to trim the vessel. Combined or separate mains for suction and discharge may be provided. Where a tank or cargo space can be used for ballast or dry cargo then either a ballast or bilge connection will be required. The system must therefore be arranged so that only the appropriate pipeline is in service; the other must be securely blanked or closed off. Where tanks are arranged for either oil or ballast a change-over chest must be fitted in the pipeline so that only the ballast main or the oil transfer main is connected to the tank.

Domestic water systems

Domestic water systems usually comprise a fresh water system for washing and drinking and a salt water system for sanitary purposes (Figure 6.21). Both use a basically similar arrangement of an automatic pump supplying the liquid to a tank which is pressurised by compressed air. The compressed air provides the head or pressure to supply the water where required. The pump is started automatically by a pressure switch which operates as the water level falls to a predetermined level. The fresh water system has, in addition, a calorifier or heater which is heated, usually with steam.

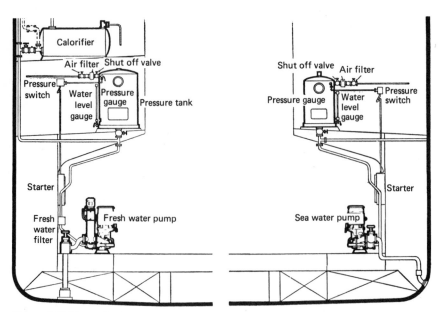

Figure 6.21 Domestic water systems

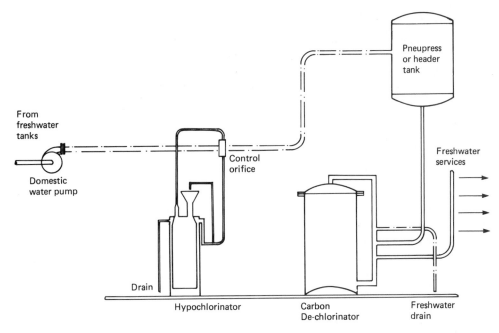

Figure 6.22 Domestic water treatment

Fresh water supplied for drinking and culinary purposes must meet purity stndards specified by the Department of Transport. Water produced from most evaporator/distillers will not meet these standards and must be treated to ensure it is biologically pure and neutral or slightly alkaline.

A treatment plant suitable for a general cargo ship is shown in Figure 6.22. The water is sterilised by an excess dose of chlorine provided as hypochlorite tablets. It is then dechlorinated in a bed of activated carbon to remove the excess chlorine. Any colour, taste and odour which was present in the water will also be removed by the carbon. Excess chlorine is originally applied to ensure that complete sterilisation occurs.

Chapter 7
Auxiliaries

Machinery, other than the main propulsion unit, is usually called 'auxiliary' even though without some auxiliaries the main machinery would not operate for long. The items considered are *air compressors, heat exchangers, distillation equipment, oil/water separators, sewage treatment plants* and *incinerators*.

Air compressor

Compressed air has many uses on board ship, ranging from diesel engine starting to the cleaning of machinery during maintenance. The air pressures of 25 bar or more are usually provided in multi-stage machines. Here the air is compressed in the first stage, cooled and compressed to a higher pressure in the next stage, and so on. The two-stage crank machine is probably the most common, and one type is shown in Figure 7.1.

Air is drawn in on the suction stroke through the first-stage suction valve via the silencer/filter. The suction valve closes on the piston upstroke and the air is compressed. The compressed air, having reached its first-stage pressure, passes through the delivery valve to the first-stage cooler. The second-stage suction and compression now take place in a similar manner, achieving a much higher pressure in the smaller, second-stage cylinder. After passing through the second-stage delivery valve, the air is again cooled and delivered to the storage system.

The machine has a rigid crankcase which provides support for the three crankshaft bearings. The cylinder block is located above and replaceable liners are fitted in the cylinder block. The running gear consists of pistons, connecting rods and the one-piece, two-throw crankshaft. The first-stage cylinder head is located on the cylinder block and the second-stage cylinder head is mounted on the first: each of the heads carries its suction and delivery valves. A chain-driven rotary-gear pump provides lubricating oil to the main bearings and through

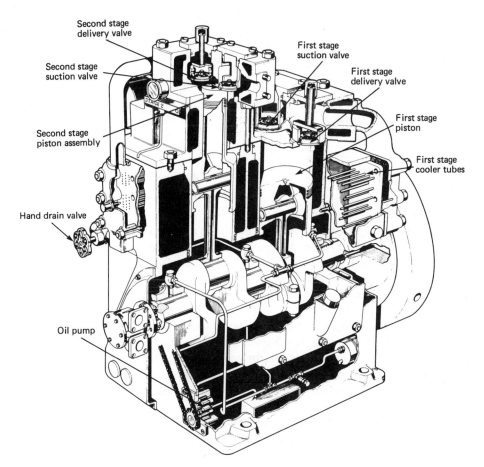

Figure 7.1 Two-stage air compressor

internally drilled passages in the crankshaft to both connecting rod
bearings. Cooling water is supplied either from an integral pump or the
machinery space system. The water passes into the cylinder block which
contains both stage coolers and then into the first and second stage
cylinder heads. A water jacket safety valve (Figure 7.2) prevents a
build-up of pressure should a cooler tube burst and compressed air
escape. Relief valves are fitted to the first and second-stage air outlets
and are designed to lift at 10% excess pressure. A fusible plug is fitted
after the second-stage cooler to limit delivered air temperature and thus
protect the compressed-air reservoirs and pipework.

Cooler drain valves are fitted to compressors. When these are open
the machine is 'unloaded' and does not produce compressed air. A
compressor when started must always be in the unloaded condition. This

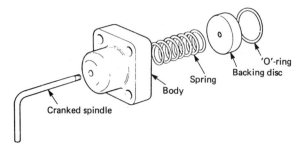

Figure 7.2 Water jacket safety valve

reduces the starting torque for the machine and clears out any accumulated moisture in the system. This moisture can affect lubrication and may produce oil/water emulsions which line the air pipelines and could lead to fires or explosions.

The compressor motor is started and the machine run up to speed. The lubricating oil pressure should be observed to build up to the correct value. The first-stage drains and then the second-stage drains are closed and the machine will begin to operate. The pressure gauge cocks should be adjusted to give a steady reading. Where manual drains are fitted they should be slightly opened to discharge any moisture which may collect in the coolers. The cooling water supply should be checked, and also operating temperatures, after a period of running loaded.

To stop the compressor, the first and second-stage cooler drain valves should be opened and the machine run unloaded for two to three minutes. This unloaded running will clear the coolers of condensate. The compressor can now be stopped and the drains should be left open. The cooling water should be isolated if the machine is to be stopped for a long period.

Automatic compressor operation is quite usual and involves certain additional equipment. An unloader must be fitted to ensure the machine starts unloaded, and once running at speed will 'load' and begin to produce compressed air. Various methods of unloading can be used but marine designs favour either depressors which hold the suction valve plates on their seats or a bypass which discharges to suction. Automatic drains must also be fitted to ensure the removal of moisture from the stage coolers. A non-return valve is usually fitted as close as possible to the discharge valve on a compressor to prevent return air flow: it is an essential fitting where unloaders are used.

The compressed air system for the supply of starting air to a diesel engine is described in Chapter 2. Control or instrument air supplies have particular requirements with regard to being moisture and oil free and without impurities. A special type of oil-free compressor may be used to supply control air or it may be treated after delivery from an ordinary air compressor. This treatment results in the air being filtered and dried in order to remove virtually all traces of oil, moisture and any atmospheric impurities.

Maintenance involves the usual checks and overhauls common to reciprocating machinery, e.g. crankcase oil level, cooling water system, operating temperatures and pressures, etc. The suction and delivery air valves for each stage will present the most work in any maintenance schedule. These valves are automatic, requiring a small pressure differential to operate.

The constant rapid opening and closing action of the valves may require the seats to be refaced. Overheating, use of incorrect lubricating oil, or the presence of dirt may result in sticking or pitting of the surfaces. The various buffer plates, spring plates, valve plate and seat which make up a suction or delivery valve can be seen in Figure 7.3. The valves should be stripped and all parts carefully cleaned and examined, any worn parts replaced and the valve seat and plate lightly lapped separately on a flat surface before reassembly to ensure a good seal.

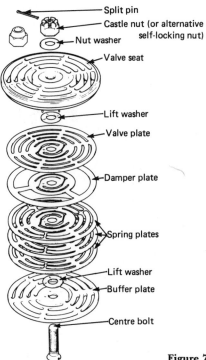

Figure 7.3 Automatic valve

Heat exchangers

Heat exchangers on board ship are mainly coolers where a hot liquid is cooled by sea water. There are some instances where liquid heating is required, such as heavy fuel oil heaters and sea water heaters for tank cleaning. Although being heat exchangers, the main condenser for a steam ship and the evaporator/distiller are dealt with separately (see Chapter 5).

The heat exchange process is accomplished by having the two liquids pass on either side of a conducting surface. The heat from the hot liquid passes to the cold liquid and the conducting surface, i.e. the tube wall, is at a temperature between the two. It is usual for marine heat exchangers to have the two liquids flowing in opposite directions, i.e. counter or contra flow. This arrangement provides a fairly constant temperature difference between the two liquids and therefore the maximum heat transfer for the available surface area.

Coolers

Coolers at sea fall into two groups, *shell and tube* and the *plate type*. Both are considered below.

Shell and tube

In the shell and tube design a tube bundle or stack is fitted into a shell (Figure 7.4). The end plates are sealed at either end of the shell and

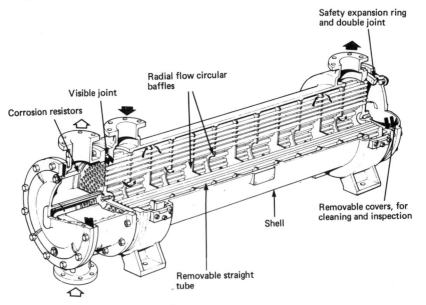

Figure 7.4 Shell and tube heat exchanger

provision is made at one end for expansion. The tubes are sealed into the tube plate at either end and provide a passageway for the cooling liquid. Headers or water boxes surround the tube plates and enclose the shell. They are arranged for either a single pass or, as in Figure 7.4, for a double pass of cooling liquid. The tube bundle has baffles fitted which serve to direct the liquid to be cooled up and down over the tubes as it passes along the cooler. The joint arrangements at the tube plate ends are different. At the fixed end, gaskets are fitted between either side of the tube plate and the shell and end cover. At the other end, the tube plate is free to move with seals fitted either side of a safety expansion ring. Should either liquid leak past the seal it will pass out of the cooler and be visible. There will be no intermixing or contamination.

Plate type

The plate-type heat exchanger is made up of a number of pressed plates surrounded by seals and held together in a frame (Figure 7.5(a)). The inlet and outlet branches for each liquid are attached to one end plate. The arrangement of seals between the plates provides passageways between adjacent plates for the cooling liquid and the hot liquid (Figure 7.5(b)). The plates have various designs of corrugations to aid heat transfer and provide support for the large, flat surface. A double seal arrangement is provided at each branch point with a drain hole to detect leakage and prevent intermixing or contamination.

Operation

Temperature control of coolers is usually achieved by adjusting the cooling liquid outlet valve. The inlet valve is left open and this ensures a constant pressure within the cooler. This is particularly important with sea water cooling where reducing pressure could lead to aeration or the collecting of air within the cooler. Air remaining in a cooler will considerably reduce the cooling effect. Vents are provided in the highest points of coolers which should be opened on first filling and occasionally afterwards. Vertical mounting of single pass coolers will ensure automatic venting. Positioning the inlet cooling water branch facing downwards and the outlet branch upwards will achieve automatic venting with horizontally mounted coolers. Drain plugs are also fitted at the lowest point in coolers.

Maintenance

Clean heat transfer surfaces are the main requirements for satisfactory operation. With sea water cooling the main problem is fouling of the surfaces, i.e. the presence of marine plant and animal growth.

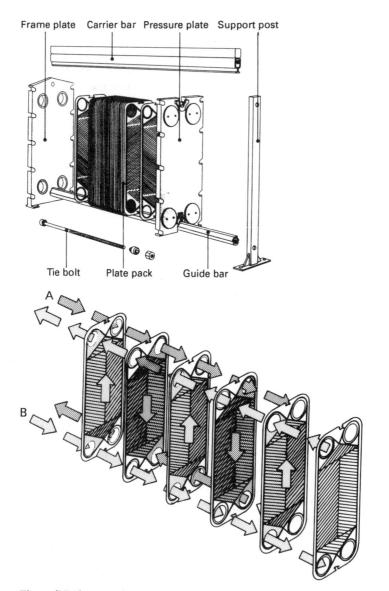

Frame plate   Carrier bar   Pressure plate   Support post

Tie bolt   Plate pack   Guide bar

A

B

Figure 7.5 Plate-type heat exchanger: (a) construction, (b) operation

With *shell and tube coolers* the end covers are removed to give access to the tubes for cleaning. Special tools are usually provided by the cooler manufacturer for cleaning the tubes. The end covers can also be cleaned.

Tube leakage can result from corrosion. This can be checked for, or identified, by having the shell side of the cooler circulated while the cooling water is shut off and the end covers removed. Any seepage into the tubes will indicate the leak. It is also possible to introduce fluorescent dyes into the shell-side liquid: any seepage will show under an ultraviolet light as a bright green glow. Leaking tubes can be temporarily plugged at each end or removed and replaced with a new tube.

Plate-type coolers which develop leaks present a more difficult problem. The plates must be visually examined to detect the faulty point. The joints between the plates can present problems in service, or on assembly of the cooler after maintenance.

Where coolers are out of use for a long period, such as during surveys or major overhauls, they should be drained on the sea water side, flushed through or washed with fresh water, and left to dry until required for service.

Heaters

Heaters, such as those used for heavy oil, are shell and tube type units, similar in construction to coolers. The heating medium in most cases is condensing steam.

Distillation systems

Distillation is the production of pure water from sea water by evaporation and re-condensing. Distilled water is produced as a result of evaporating sea water either by a boiling or a flash process. This evaporation enables the reduction of the 32 000 parts per million of dissolved solids in sea water down to the one or two present in distilled water. The machine used is called an 'evaporator', although the word 'distiller' is also used.

Boiling process

Sea water is boiled using energy from a heating coil, and by reducing the pressure in the evaporator shell, boiling can take place at about 60°C.

The sea water from the ship's services is first circulated through the condenser and then part of the outlet is provided as feed to the evaporation chamber (Figure 7.6). Hot diesel engine jacket water or steam is passed through the heater nest and, because of the reduced pressure in the chamber, the sea water boils. The steam produced rises and passes through a water separator or demister which prevents water droplets passing through. In the condensing section the steam becomes pure water, which is drawn off by a distillate pump. The sea water feed is regulated by a flow controller and about half the feed is evaporated. The

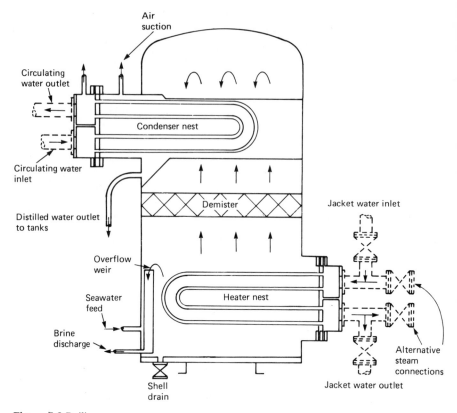

Figure 7.6 Boiling process evaporator

remainder constantly overflows a weir and carries away the extra salty water or brine. A combined brine and air ejector draws out the air and brine from the evaporator.

Flash process

Flash evaporation is the result of a liquid containing a reasonable amount of sensible heat at a particular pressure being admitted to a chamber at a lower pressure. The liquid immediately changes into steam, i.e. it flashes, without boiling taking place. The sensible heat content, water pressure and chamber pressure are designed to provide a desired rate of evaporation. More than one stage of evaporation can take place by admitting the liquid into chambers with progressively lower pressures.

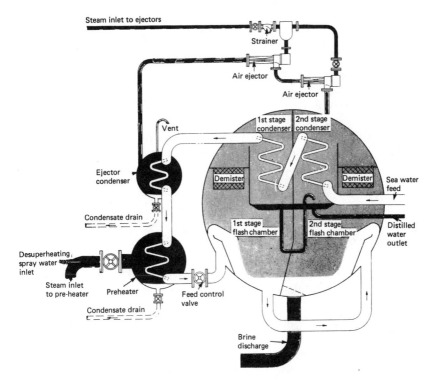

Figure 7.7 Two-stage flash evaporator

A two-stage flash evaporator is shown in Figure 7.7. The feed pump circulates sea water through the vapour condensers and the preheater. The heated sea water then passes to the first-stage flash chamber where some of it flashes off. A demister removes any water droplets from the steam as it rises and is then condensed in the first-stage condenser.

The heated sea water passes to the second-stage flash chamber, which is at a lower pressure, and more water flashes off. This steam is demisted and condensed and, together with the distilled water from the first-stage, is drawn off by the distillate pump.

The concentrated sea water or brine remaining in the second-stage flash chamber is drawn off by the brine pump. The preheater uses steam to heat the sea water and most of the latent heat from the flash steam is returned to the sea water passing through the condensers. An air ejector is used to maintain the low pressure in the chambers and to remove any gases released from the sea water.

Maintenance

During the operation of evaporating plants, scale will form on the heating surfaces. The rate of scale formation will depend upon the operating temperature, the flow rate and density of the brine.

Scale formation will result in greater requirements for heating to produce the rated quantities of distilled water or a fall-off in production for a fixed heating supply.

Cold shocking, the alternate rapid heating and cooling of the tube surfaces, for a boiling process type, can reduce scale build-up. Ultimately, however, the plant must be shut down and the scale removed either by chemical treatment or manual cleaning.

Oil/water separators

Oil/water separators are used to ensure that ships do not discharge oil when pumping out bilges, oil tanks or any oil-contaminated space. International legislation relating to oil pollution is becoming more and more stringent in the limits set for oil discharge. Clean water suitable for discharge is defined as that containing less than 15 parts per million of oil. Oil/water separators using the gravity system can only achieve 100 parts per million and must therefore be used in conjunction with some form of filter. Depending upon the size of ship a discharge purity of 100 or 15 parts per million will be required. Where 100 parts per million purity is required the oil/water separator may be used alone.

A complete oil/water separator and filter unit for 15 parts per million purity is shown in Figure 7.8. The complete unit is first filled with clean water; the oily water mixture is then pumped through the separator inlet pipe into the coarse separating compartment. Here some oil, as a result of its lower density, will separate and rise into the oil collection space. The remaining oil/water mixture now flows down into the fine separating compartment and moves slowly between the catch plates. More oil will separate out onto the underside of these plates and travel outwards until it is free to rise into the oil collecting space. The almost oil-free water passes into the central pipe and leaves the separator unit. The purity at this point will be 100 parts per million or less. An automatically controlled valve releases the separated oil to a storage tank. Air is released from the unit by a vent valve. Steam or electric heating coils are provided in the upper and sometimes the lower parts of the separator, depending upon the type of oil to be separated.

Where greater purity is required, the almost oil-free water passes to a filter unit. The water flows in turn through two filter stages and the oil removed passes to oil collecting spaces. The first-stage filter removes

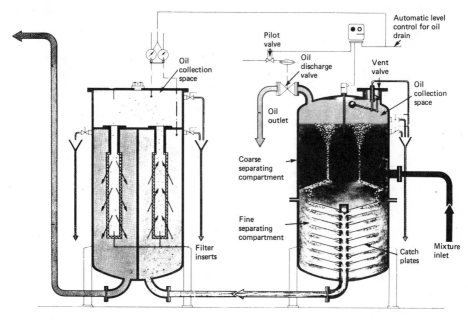

Figure 7.8 Oily water separator

physical impurities present and promotes some fine separation. The second-stage filter uses coalescer inserts to achieve the final de-oiling. Coalescence is the breakdown of surface tension between oil droplets in an oil/water mixture which causes them to join and increase in size. The oil from the collecting spaces is drained away manually, as required, usually about once a week. The filter inserts will require changing, the period of useful life depending upon the operating conditions.

The latest legislative requirements are, where 100 parts per million purity is required, a monitoring unit which continuously records, and, where 15 parts per million purity is necessary, an alarm unit to provide warning of levels of discharge in excess of 15 parts per million.

Sewage treatment

The discharge of untreated sewage in controlled or territorial waters is usually banned by legislation. International legislation is in force to cover any sewage discharges within specified distances from land. As a result, and in order to meet certain standards all new ships have sewage treatment plants installed.

Untreated sewage as a suspended solid is unsightly. In order to break down naturally, raw sewage must absorb oxygen. In excessive amounts it could reduce the oxygen content of the water to the point where fish and plant life would die. Pungent smells are also associated with sewage as a result of bacteria which produce hydrogen sulphide gas. Particular bacteria present in the human intestine known as *E. coli* are also to be found in sewage. The *E. coli* count in a measured sample of water indicates the amount of sewage present.

Two particular types of sewage treatment plant are in use, employing either chemical or biological methods. The chemical method is basically a storage tank which collects solid material for disposal in permitted areas or to a shore collection facility. The biological method treats the sewage so that it is acceptable for discharge inshore.

Chemical sewage treatment

This system minimises the collected sewage, treats it and retains it until it can be discharged in a decontrolled area, usually well out to sea. Shore receiving facilities may be available in some ports to take this retained sewage.

This system must therefore collect and store sewage produced while the ship is in a controlled area. The liquid content of the system is reduced, where legislation permits, by discharging wash basins, bath and shower drains straight overboard. Any liquid from water closets is treated and used as flushing water for toilets. The liquid must be treated such that it is acceptable in terms of smell and appearance.

A treatment plant is shown diagrammatically in Figure 7.9. Various chemicals are added at different points for odour and colour removal and also to assist breakdown and sterilisation. A comminutor is used to physically break up the sewage and assist the chemical breakdown process. Solid material settles out in the tank and is stored prior to discharge into the sullage tank: the liquid is recycled for flushing use.

Tests must be performed daily to check the chemical dosage rates. This is to prevent odours developing and also to avoid corrosion as a result of high levels of alkalinity.

Biological sewage treatment

The biological system utilises bacteria to completely break down the sewage into an acceptable substance for discharge into any waters. The extended aeration process provides a climate in which oxygen-loving bacteria multiply and digest the sewage, converting it into a sludge. These oxygen-loving bacteria are known as *aerobic*.

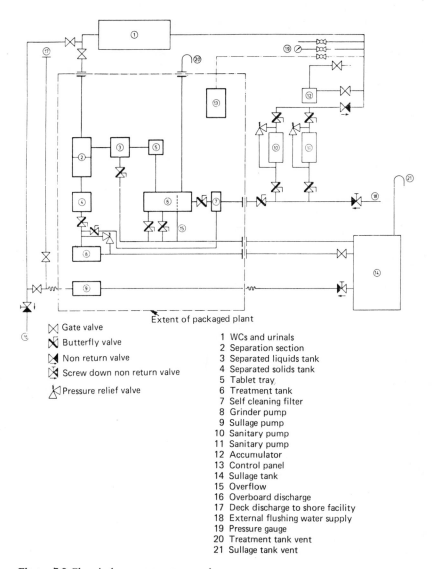

Gate valve
Butterfly valve
Non return valve
Screw down non return valve
Pressure relief valve

1 WCs and urinals
2 Separation section
3 Separated liquids tank
4 Separated solids tank
5 Tablet tray
6 Treatment tank
7 Self cleaning filter
8 Grinder pump
9 Sullage pump
10 Sanitary pump
11 Sanitary pump
12 Accumulator
13 Control panel
14 Sullage tank
15 Overflow
16 Overboard discharge
17 Deck discharge to shore facility
18 External flushing water supply
19 Pressure gauge
20 Treatment tank vent
21 Sullage tank vent

Figure 7.9 Chemical sewage treatment plant

The treatment plant uses a tank which is divided into three watertight compartments: an aeration compartment, settling compartment and a chlorine contact compartment (Figure 7.10). The sewage enters the aeration compartment where it is digested by aerobic bacteria and micro-organisms, whose existence is aided by atmospheric oxygen which is pumped in. The sewage then flows into the settling compartment where the activated sludge is settled out. The clear liquid flows to the

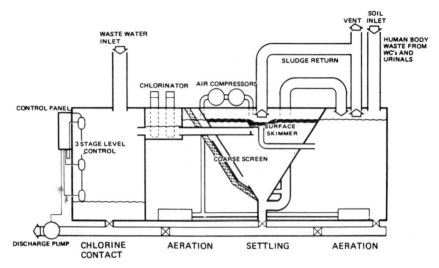

Figure 7.10 Biological sewage treatment plant

chlorinator and after treatment to kill any remaining bacteria it is discharged. Tablets are placed in the chlorinator and require replacement as they are used up. The activated sludge in the settling tank is continuously recycled and builds up, so that every two to three months it must be partially removed. This sludge must be discharged only in a decontrolled area.

Incinerator

Stricter legislation with regard to pollution of the sea, limits and, in some instances, completely bans the discharge of untreated waste water, sewage, waste oil and sludge. The ultimate situation of no discharge can be achieved by the use of a suitable incinerator. When used in conjunction with a sewage plant and with facilities for burning oil sludges, the incinerator forms a complete waste disposal package.

One type of incinerator for shipboard use is shown in Figure 7.11. The combustion chamber is a vertical cylinder lined with refractory material. An auxiliary oil-fired burner is used to ignite the refuse and oil sludge and is thermostatically controlled to minimise fuel consumption. A sludge burner is used to dispose of oil sludge, water and sewage sludge and works in conjunction with the auxiliary burner. Combustion air is provided by a forced draught fan and swirls upwards from tangential ports in the base. A rotating-arm device accelerates combustion and also

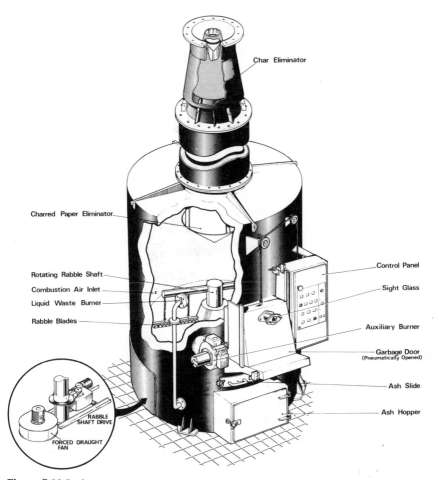

Char Eliminator

Charred Paper Eliminator

Rotating Rabble Shaft
Combustion Air Inlet
Liquid Waste Burner

Rabble Blades

Control Panel

Sight Glass

Auxiliary Burner

Garbage Door
(Pneumatically Opened)

Ash Slide

Ash Hopper

RABBLE
SHAFT DRIVE

FORCED DRAUGHT
FAN

Figure 7.11 Incinerator

clears ash and non-combustible matter into an ash hopper. The loading door is interlocked to stop the fan and burner when opened.

Solid material, usually in sacks, is burnt by an automatic cycle of operation. Liquid waste is stored in a tank, heated and then pumped to the sludge burner where it is burnt in an automatic cycle. After use the ash box can be emptied overboard.

To Finish

Chapter 8

Fuel oils, lubricating oils and their treatment

Crude oil is, at the present time, the source of most fuel oils for marine use. Synthetic fuels are being developed but will probably be too expensive for ship propulsion. Solid fuel, such as coal, is returning in a small way for certain specialised trade runs. The various refined products of crude oil seem likely to remain as the major forms of marine fuel.

The refining process for crude oil separates by heating and distillation the various fractions of the oil. Paraffin fuel would be used in gas turbine plants, gas oil in high- and medium-speed diesel engines and crude oils in slow-speed and some medium-speed diesels. Paraffin and gas oil are known as 'distillates', which are free flowing, easily stored and can be used without further treatment. Residual fuels, however, are very viscous or thick at normal temperatures, and require heating before use. Additional treatment to remove harmful chemicals or sulphur may be required for all or some of the refined products, depending upon their application. Finally blending or mixing of the various oils is done to provide a range of commercial fuels for different duties.

Fuel oils

Fuel oils have various properties which determine their performance and are quoted in specifications. The *specific gravity* or *relative density* is the weight of a given volume of fuel compared to the weight of the same volume of water expressed as a ratio, and measured at a fixed temperature. *Viscosity* is a resistance to flow. A highly viscous fuel will therefore require heating in order to make it flow. Measurement of viscosity is by Redwood, Saybolt or Engler instrument flow times for a given volume of fuel.

The *ignition quality* of a fuel is measured by the time delay between injection and combustion, which should be short for good controlled burning. Ignition quality is indicated as cetane number, diesel index and calculated cetane index; the higher the value the better the ignition quality of the fuel.

The *flash point* is a figure obtained and used mainly to indicate the maximum safe storage temperature. The test determines the temperature at which the fuel will give off sufficient vapours to ignite when a flame is applied. Two values are possible: an open flash point for atmospheric heating, and a closed flash point when the fuel is covered while heating.

Low-temperature properties are measured in terms of *pour point* and *cloud point*. The pour point is slightly above the temperature at which the fuel just flows under its own weight. It is the lowest temperature at which the fuel can be easily handled. At the cloud point waxes will form in the fuel. Below the cloud point temperature, pipe or filter blocking may occur.

The carbon residue forming property of a fuel is usually measured by the Conradson method. Controlled burning of a fuel sample gives a measure of the *residual carbon and other remains*.

Sulphur content is of importance since it is considered a cause of engine wear. A maximum limit, expressed as a percentage by weight, is usually included in specifications.

The *calorific value* of a fuel is the heat energy released during combustion. Two values are used, the more common being the Higher Calorific Value, which is the heat energy resulting from combustion. The Lower Calorific Value is a measure of the heat energy available and does not include the heat energy contained in steam produced during combustion but passing away as exhaust. The measurement is obtained from a bomb calorimeter test where a small fuel quantity is burnt under controlled conditions.

The various fuel properties have different effects on performance of the engine and the storage and handling requirements of the system. Blending and the use of various additives will also influence both the engine and the system.

Viscosity will affect jerk-type injector pumps and injector operation since the liquid fuel is the operating medium. The pump mechanism is lubricated by the fuel which, if it is of low viscosity, will cause wear.

Cloud point and pour point values are important when considering the lowest system operating temperatures. Wax deposited in filters and fuel lines will cause blockages and may restrict fuel flow to the engine.

The cetane number or diesel index will determine injection timing and also influences the combustion noise and production of black smoke. The temperature in a fuel system should be progressively increased in order to deliver fuel at the correct viscosity to the injectors or burners. System cleanliness is also very important to reduce wear on the many finely machined parts in the fuel injection equipment. Regular attention to filters and general system cleanliness is essential. Various additives are used to, for instance, remove lacquer from metal surfaces, reduce wear and prevent rust.

Lubricating oils

Lubricating oils are a product of the crude oil refining process. The various properties required of the oil are obtained as a result of blending and the introduction of additives. The physical and chemical properties of an oil are changed by additives which may act as oxidation inhibitors, wear reducers, dispersants, detergents, etc. The important lubricant properties will now be examined.

Viscosity has already been mentioned with respect to fuel oils, but it is also an important property of lubricating oils. Viscosity index is also used, which is the rate of change of viscosity with temperature.

The *Total Base Number* (TBN) is an indication of the quantity of alkali, i.e. base, which is available in a lubricating oil to neutralise acids.

The *acidity* of an oil must be monitored to avoid machinery damage and neutralisation number is used as the unit of measurement.

The *oxidation resistance* of a lubricant can also be measured by neutralisation number. When excessively oxidised an oil must be discarded.

The *carbon-forming tendency* of a lubricating oil must be known, particularly for oils exposed to heat. A carbon residue test is usually performed to obtain a percentage value.

The *demulsibility* of an oil refers to its ability to mix with water and then release the water in a centrifuge. This property is also related to the tendency to form sludge.

Corrosion inhibition relates to the oil's ability to protect a surface when water is present in the oil. This is important where oils can be contaminated by fresh or salt water leaks.

The modern lubricant must be capable of performing numerous duties. This is achieved through blending and additives. It must prevent metal-to-metal contact and reduce friction and wear at moving parts. The oil must be stable and not break down or form carbon when exposed to high temperatures, such as where oil cooling is used. Any contaminants, such as acidic products of combustion, must be neutralised by alkaline additives; any carbon build up on surfaces must be washed away by detergent additives and held in suspension by a dispersant additive. The oil must also be able to absorb water and then release it during purification, but meanwhile still protect the metal parts from corrosion.

The various types of engine and other equipment will have oils developed to meet their particular duties.

Trunk piston engine lubricating oil must lubricate the cylinders as well as the crankcase: some contamination from the products of combustion will therefore occur, resulting in acidity and carbon deposits. The oil must, in addition to lubricating, neutralise the acids and absorb the deposits.

Turbine oil, while lubricating the moving parts, must also carry away considerable quantities of heat from the bearings. This calls for a stable oil which will not break down at high temperatures or form deposits. Where gearbox lubrication is also required certain extreme pressure (EP) additives will be needed to assist the lubricating film. Contact with water in the form of steam will be inevitable so good demulsifying properties will be essential.

Slow-speed diesel engines will have separate cylinder and crankcase lubrication systems. The cylinder oil will have to neutralise the acidic products of combustion and also have good detergent properties to keep the metal surfaces clean. Crankcase oils are either detergent type, multi-purpose oils or rust and oxidation inhibited. Good demulsification and anti-corrosive properties are required together with oxidation resistance which is provided by the inhibited crankcase oil. The detergent or multi-purpose oil is particularly useful where oil cooling of pistons occurs or where contamination by combustion products is possible.

Oil treatment

Both fuel oils and lubricating oils require treatment before passing to the engine. This will involve storage and heating to allow separation of water present, coarse and fine filtering to remove solid particles and also centrifuging.

The centrifugal separator is used to separate two liquids, for example oil and water, or a liquid and solids as in contaminated oil. Separation is speeded up by the use of a centrifuge and can be arranged as a continuous process. Where a centrifuge is arranged to separate two liquids, it is known as a 'purifier'. Where a centrifuge is arranged to separate impurities and small amounts of water from oil it is known as a 'clarifier'.

The separation of impurities and water from fuel oil is essential for good combustion. The removal of contaminating impurities from lubricating oil will reduce engine wear and possible breakdowns. The centrifuging of all but the most pure clean oils is therefore an absolute necessity.

Centrifuging

A centrifuge consists of an electric motor drive to a vertical shaft on the top of which is mounted the bowl assembly. An outer framework surrounds the assembly and carries the various feed and discharge connections. The bowl can be a solid assembly which retains the separated sludge and operates non-continuously, or the bowl can be

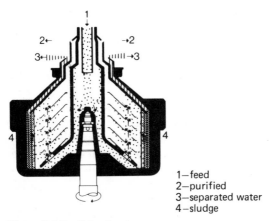

1—feed
2—purified
3—separated water
4—sludge

Figure 8.1 Purifying bowl arrangement

arranged so that the upper and lower parts separate and the sludge can be discharged while the centrifuge operates continuously. The dirty oil is admitted into the centre of the bowl, passes up through a stack of discs and out through the top (Figure 8.1).

The purifying process

The centrifugal separation of two liquids, such as oil and water, results in the formation of a cylindrical interface between the two. The positioning of this interface within the centrifuge is very important for correct operation. The setting or positioning of the interface is achieved by the use of dam rings or gravity discs at the outlet of the centrifuge. Various diameter rings are available for each machine when different densities of oil are used. As a general rule, the largest diameter ring which does not break the 'seal' should be used.

The clarifying process

Cleaning oil which contains little or no water is achieved in a clarifier bowl where the impurities and water collect at the bowl periphery. A

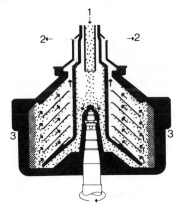

Figure 8.2 Clarifying bowl arrangement

clarifier bowl has only one outlet (Figure 8.2). No gravity disc is necessary since no interface is formed; the bowl therefore operates at maximum separating efficiency since the oil is subjected to the maximum centrifugal force.

The bowl discs

Purifier and clarifier bowls each contain a stack of conical discs. The discs may number up to 150 and are separated from one another by a small gap. Separation of impurities and water from the oil takes place between these discs. A series of aligned holes near the outside edge permits entry of the dirty oil. The action of centrifugal force causes the lighter components (the clean oil) to flow inwards and the water and impurities flow outwards. The water and impurities form a sludge which moves outwards along the undersides of the discs to the periphery of the bowl.

Non-continuous operation

Certain designs of centrifuges are arranged for a short period of operation and are then shut down for cleaning. After cleaning and removal of the sludge from the bowl, the machine is returned to service. Two different designs are used for this method of operation: a long narrow bowl and a short wide bowl. The narrow-bowl machine has to be cleaned after a shorter running period and requires dismantling in order to clean the bowl. Cleaning of the bowl is, however, much simpler since it does not contain a stack of discs. The wide-bowl machine can be cleaned in place, although there is the added complication of the stack of conical discs which must be cleaned.

Continuous operation

Modern wide-bowl centrifuge designs enable continuous operation over a considerable period of time. This is achieved by an ejection process which is timed to discharge the sludge at regular intervals. The sludge deposits build up on the bowl periphery as separation continues, and the ejection process is timed to clear these deposits before they begin to affect the separation process. To commence the ejection process the oil feed to the centrifuge is first shut off and the oil remaining in the bowl is removed by admitting flushing water. Water is then fed into the hydraulic system in the bottom of the bowl to open a number of spring-loaded valves. This 'operating' water causes the sliding bowl bottom to move downwards and open discharge ports in the bowl periphery. The sludge is discharged through these ports by centrifugal

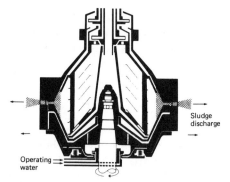

Figure 8.3 Sludge discharge

force (Figure 8.3). Closing 'operating' water is now fed in to raise the sliding bowl up again and close the discharge ports. Water is fed into the bowl to remake the liquid seal required for the separation process, the oil feed reopened, and separation continues.

The complete ejection cycle takes only a few seconds and the centrifuge is in continuous operation throughout. Different bowl designs exist for various forms of sludge discharge, e.g. total discharge, controlled partial discharge, and so on. With controlled partial discharge the oil supply is not shut off and all of the sludge is discharged. In this way the separation process is not stopped. Whatever method is adopted the centrifuge can be arranged so that the discharge process is performed manually or by an automatic timer.

Maintenance

The bowl and the disc stack will require periodical cleaning whether or not an ejection process is in operation. Care should be taken in stripping down the bowl, using only the special tools provided and noting that some left-hand threads are used. The centrifuge is a perfectly balanced piece of equipment, rotating at high speeds: all parts should therefore be handled and treated with care.

Heavy fuel oil separation

Changes in refinery techniques are resulting in heavy fuel oils with increased density and usually contaminated with catalytic fines. These are small particles of the catalysts used in the refining process. They are extremely abrasive and must be removed from the fuel before it enters the engine. The generally accepted maximum density limit for purifier operation is 991 kg/m^3 at 15°C.

In the ALCAP separation system the separator has no gravity disc and operates, to some extent, as a clarifier. Clean oil is discharged from the oil outlet and separated sludge and water collect at the periphery of the bowl. When the separated water reaches the disc stack, some water will escape with the cleaned oil. The increase in water content is sensed by a water-detecting transducer in the outlet (Figure 8.4). The water

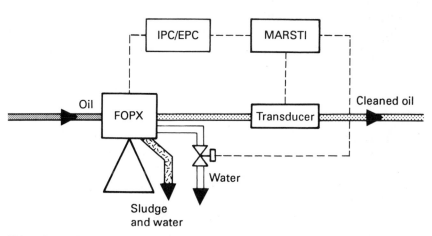

Figure 8.4 Fuel oil separation control

transducer signal is fed to the MARST 1 microprocessor which will discharge the water when a predetermined level is reached. The water will be discharged from sludge ports in the bowl or, if the amount is large, from a water drain valve.

The ALCAP system has also proved effective in the removal of catalytic fines from fuel oil.

Lubricating oil centrifuging

Diesel engines

Lubricating oil in its passage through a diesel engine will become contaminated by wear particles, combustion products and water. The centrifuge, arranged as a purifier, is used to continuously remove these impurities.

The large quantity of oil flowing through a system means that full flow lubrication would be too costly. A bypass system, drawing dirty oil from low down in the oil sump remote from the pump suction and returning clean oil close to the pump suction, is therefore used. Since this is a bypass system the aim should be to give the lowest degree of impurity for the complete system, which may mean running the centrifuge somewhat below the maximum throughput.

Water-washing during centrifuging can be adopted if the oil manufacturer or supplier is in agreement; but some oils contain water-soluble additives, which would of course be lost if water-washed. The advantages of water-washing include the dissolving and removal of acids, improved separation by wetting solid impurities, and the constant renewal of the bowl liquid seal. The washing water is usually heated to a slightly higher temperature than the oil.

Detergent-type oils are used for cleaning as well as lubricating and find a particular application in trunk-type engines and some slow-speed engines. Detergent-type oil additives are usually soluble and must not therefore be water-washed.

Steam turbines

The lubricating oil in a steam turbine will become contaminated by system impurities and water from condensed steam, so bypass separation is used to clean the oil. The dirty oil is drawn from the bottom of the sump and clean oil returned near the pump suction. Preheating of the oil before centrifuging assists the separation process. Water washing of the oil can be done where the manufacturer or supplier of the oil permits it.

Homogenisers

A homogeniser is used to create a stable oil and water emulsion which can be burnt in a boiler or diesel engine. Such an emulsion is considered to bring about more efficient combustion and also reduce solid emissions in the exhaust gas.

Various designs utilise an impact or rolling action to break down the fuel particles and mix them with the water. It is also considered that agglomerates of asphaltenes and bituminous matter are broken down and can therefore be burnt. The manufacturers contend that a homogeniser can render a sludge burnable whereas a centrifuge would remove such material. Homogenisers are able to reduce catalytic fines into finely ground particles which will do no harm.

Shipboard experience with homogenisers is limited and generally not favourable. Most authorities consider it better to remove water and solid contaminants than simply grind them down.

Blenders

Blending is the mixing of two fuels, usually a heavy fuel and marine diesel oil. The intention is to produce an intermediate-viscosity fuel suitable for use in auxiliary diesels. The fuel cost savings for intermediate fuel grades are sufficient to justify the cost of the blending plant. Furthermore no supply problems exist since the appropriate mixture can be produced by the blender from available heavy and marine diesel oils.

The blending unit thoroughly mixes the two fuels in the appropriate proportions before supplying it to a blended fuel supply tank.

Compatibility can be a problem and tests should be conducted on any new fuel before it is used. Incompatible fuels may produce sludge or sediment. The cracked residues presently supplied from many refineries are very prone to incompatibility problems when blended with marine diesel oil.

Filters and strainers

Mechanical separation of solid contaminants from oil systems (fuel and lubricating) is achieved by the use of filters and strainers. A strainer is usually a coarse filter to remove the larger contaminating particles. Both are arranged as full flow units, usually mounted in pairs (duplex) with one as a standby.

The strainer usually employs a mesh screen, an assembly of closely packed metal plates or wire coils which effectively block all but the smallest particles. It is usually fitted on the suction side of a pump and must be cleaned regularly or when the pressure differential across it become unacceptable. Where suction conditions are critical the strainer will be fitted on the discharge side of the pump. When cleaning is undertaken the other unit will be connected into the system by changeover valves or levers and oil circulation will continue. The particles of dirt collect on the outside of the strainer element or basket and can be removed by compressed air or brushing. A strainer should be cleaned as soon as it is taken out of the system, then reassembled and left ready for use.

Magnetic strainers are often used in lubricating oil systems, where a large permanent magnet collects any ferrous particles which are circulating in the system. The magnet is surrounded by a cage or basket to simplify cleaning.

Fine filters, again in pairs, are used to remove the smallest particles of dirt from oil before the oil enters the finely machined engine parts in either the fuel injection system or the bearings of the rotating machinery. Fine filters are full-flow units which clean all the oil supplied to the engine. The filtering substance may be a natural or synthetic fibrous woollen felt or paper. A felt-type fine filter is shown in Figure 8.5. A steel division plate divides the steel pressure vessel into an upper and a lower chamber. Dirty oil passes into the upper chamber and through the filter element, then the filtered oil passes down the central tube to the lower chamber and out of the unit. A magnetic filter can be

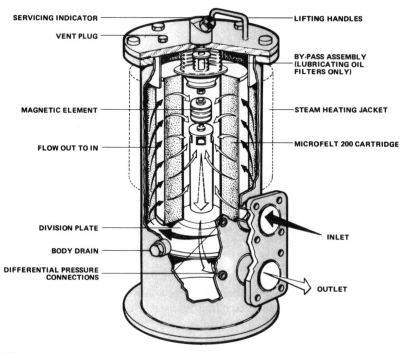

Figure 8.5 Fine filter

positioned as shown in the central tube. A spring-loaded bypass is shown in the diagram, for lubricating oil filters only, to ensure a flow of oil should the filter become blocked. The cartridge in the design shown is disposable although designs exist to enable back-flushing with compressed air to clean the filter element as required. The filter unit shown will be one of a pair which can be alternately in service.

In full-flow filtration systems all the oil passes through the filter on its way to the engine. In a by-pass system most of the oil goes to the lubrication system and a part is by-passed to a filter. A higher pressure drop across the filter can then be used and a slower filtration rate. A centrifugal filter can be used in a by-pass system where the oil passes through a rotor and spins it at high speed (Figure 8.6). Dirt particles in the oil are then deposited on the walls of the rotor and the clean oil

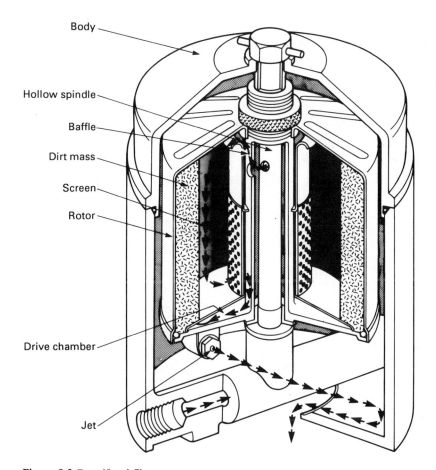

Figure 8.6 Centrifugal filter

returns to the sump. This type of filter cannot block or clog and requires no replaceable elements. It must be dismantled for cleaning of the rotor unit at regular intervals.

Microbiological infestation

Minute microorganisms, i.e. bacteria, can exist in lubricating oils and fuel oils. Under suitable conditions they can grow and multiply at phenomenal rates. Their presence leads to the formation of acids and sludge, metal staining, deposits and serious corrosion. The presence of slime and the smell of rotten eggs (hydrogen sulphide) indicates a contaminated system.

Water in a lubricating oil or fuel oil, oxygen and appropriate temperature conditions will result in the growth of bacteria and infestation of a system. The removal of water, or ensuring its presence is at a minimum, is the best method of infestation prevention. The higher the temperature in settling, service and drain tanks holding fuel or lubricating oils, the better.

Test kits are available to detect the presence of bacteria, and biocides can be used to kill all bacteria present in a system. The system must then be thoroughly flushed out.

Chapter 9
Refrigeration, air conditioning and ventilation

Refrigeration is a process in which the temperature of a space or its contents is reduced to below that of their surroundings. *Air conditioning* is the control of temperature and humidity in a space together with the circulation, filtering and refreshing of the air. *Ventilation* is the circulation and refreshing of the air in a space without necessarily a change of temperature. With the exception of special processes, such as fish freezing, air is normally employed as the heat transfer medium. As a result fans and ducting are used for refrigeration, air conditioning and ventilation. The three processes are thus interlinked and all involve the provision of a suitable climate for men, machinery and cargo.

Refrigeration

Refrigeration of cargo spaces and storerooms employs a system of components to remove heat from the space being cooled. This heat is transferred to another body at a lower temperature. The cooling of air for air conditioning entails a similar process.

The transfer of heat takes place in a simple system: firstly, in the evaporator where the lower temperature of the refrigerant cools the body of the space being cooled; and secondly, in the condenser where the refrigerant is cooled by air or water. The usual system employed for marine refrigeration plants is the vapour compression cycle, for which the basic diagram is shown in Figure 9.1.

The pressure of the refrigerant gas is increased in the compressor and it thereby becomes hot. This hot, high-pressure gas is passed through into a condenser. Depending on the particular application, the refrigerant gas will be cooled either by air or water, and because it is still at a high pressure it will condense. The liquid refrigerant is then distributed through a pipe network until it reaches a control valve alongside an evaporator where the cooling is required. This regulating valve meters the flow of liquid refrigerant into the evaporator, which is

163

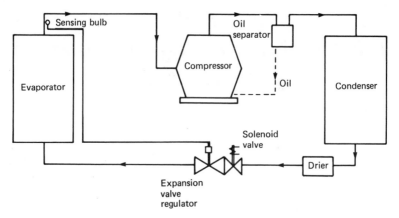

Figure 9.1 Vapour compression cycle

at a lower pressure. Air from the cooled space or air conditioning system is passed over the evaporator and boils off the liquid refrigerant, at the same time cooling the air. The design of the system and evaporator should be such that all the liquid refrigerant is boiled off and the gas slightly superheated before it returns to the compressor at a low pressure to be recompressed.

Thus it will be seen that heat that is transferred from the air to the evaporator is then pumped round the system until it reaches the condenser where it is transferred or rejected to the ambient air or water.

It should be noted that where an air-cooled condenser is employed in very small plants, such as provision storerooms, adequate ventilation is required to help remove the heat being rejected by the condenser. Also, in the case of water-cooled condensers, fresh water or sea water may be employed. Fresh water is usual when a central fresh-water/sea-water heat exchanger is employed for all engine room requirements. Where this is the case, because of the higher cooling-water temperature to the condenser, delivery temperatures from condensers will be higher than that on a sea water cooling system.

Refrigerants

Generally speaking these are sub-divided into primary and secondary refrigerants.

Primary refrigerants

This is the refrigerant employed in the compressor, condenser and evaporator system and certain properties are essential requirements. For example it will boil off or evaporate at a low temperature and reasonable pressure and it will condense at a temperature near normal sea water temperature at a reasonable pressure. The refrigerant must also be free from toxic, explosive, flammable and corrosive properties where possible. Some refrigerants have critical temperatures above which the refrigerant gas will not condense. This was one of the disadvantages of carbon dioxide, which was used for many years on ships. Ships operating in areas with very high sea-water temperatures had difficulty in liquefying the carbon dioxide without some additional sub-cooling system. A further disadvantage of carbon dioxide was the very high pressure at which the system operated, resulting in large and heavy machinery.

Between the carbon dioxide era and the present refrigerants, methyl chloride and ammonia were used. Due to its explosive properties, methyl chloride is now banned for shipboard use. Ammonia is still employed, but requires special ventilation.

The modern refrigerants are fluorinated hydrocarbon compounds of various formulae, with the exception of Refrigerant 502, which is an azeotropic (fixed boiling point) mixture of Refrigerant 22 and Refrigerant 115. These are usually refered to as 'Freons' with a number related to their particular formula.

Refrigerant 11 is a very low-pressure refrigerant which requires a large circulation for a particular cooling effect. It has particular advantages when used in air conditioning units, since it will have a low power consumption.

Refrigerant 12 was one of the first fluorinated hydrocarbon refrigerants, as these numbered substances are known, to become readily and cheaply available. A disadvantage is that evaporator pressures are below atmospheric and any system leaks draw in air and moisture.

Refrigerant 22 is now probably the most common refrigerant. It provides a considerable range of low-temperature operation before the evaporator pressure drops below atmospheric conditions. There is also a space saving as the compressor displacement is about 60% of that required for Refrigerant 12.

Refrigerant 502's particular advantages are that the displacement required is similar to that of Refrigerant 22. Gas delivery temperatures from the compressor are greatly reduced, and therefore there is less likely to be a break-up of the lubricating oil and stressing of the delivery valves.

All the above refrigerants are non-corrosive and may be used in hermetic or semi-hermetic compressor units. Refrigerant 502, however, does have less effect on the lacquers and elastomers employed in compressors and motors. At present Refrigerant 502 is still an expensive gas and not readily available worldwide.

Secondary refrigerants

Both large air conditioning and cargo cooling systems may employ a secondary refrigerant. In this case the primary refrigerant evaporator will be circulated with the secondary refrigerant, which is then passed to the space to be cooled. Secondary refrigerants are employed where the installation is large and complex to avoid the circulation of expensive primary refrigerants in large quantities. These primary refrigerants can be very searching, that is they can escape through minute clearances, so it is essential to keep the number of possible leakage points to a minimum.

In the case of air conditioning plants, fresh water is the normal secondary refrigerant, which may or may not have a glycol solution added. The more common secondary refrigerant on large cargo installations is a calcium chloride brine to which is added inhibitors to prevent corrosion.

System components

Compressors

There are three types of compressor in use at sea: centrifugal, reciprocating, and screw.

Centrifugal compressors are used with Refrigerants 11 or 12 and are limited in their application to large air conditioning installations. They are similar in appearance to horizontal centrifugal pumps and may have one or more stages.

Reciprocating compressors cover the whole spectrum of refrigeration requirements at sea, from air conditioning to low temperature cargo installations. They are normally of a compact design and may be of an in-line, V or W configuration. Figure 9.2 shows a 4-cylinder W configuration. The construction arrangement can be seen and the principle of operation is similar in many respects to an air compressor. For low-temperature applications the machine may be arranged as a two-stage compressor and some machines are made so that they can be changed from single to two stage, depending on cargo requirements. As the crankcase is subject to refrigerant pressure, the drive shaft seal is required to prevent a flow of refrigerant out of the compressor or ingress of air. In semi-hermetic or hermetic machines this problem is obviated as the motor and compressor are in one casing.

Screw compressors have replaced reciprocating compressors in large installations for two reasons. Firstly, fewer and more compact machines are used; secondly, a reduced number of working parts results in greater reliability with reduced maintenance requirements. There are two types of screw compressor; one employs two rotors side by side and the other,

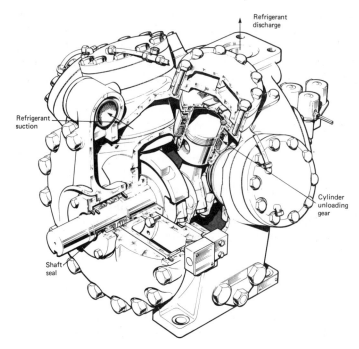

Figure 9.2 Reciprocating compressor

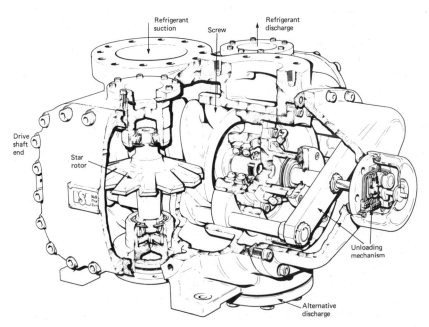

Figure 9.3 Single-screw compressor

which is a more modern development, is a single rotor with two star wheels, one on either side. As the star wheels compress the gas in opposite directions, the thrust on this type of rotor is balanced. Such a compressor in shown in Figure 9.3. The principle of operation for both types is similar to a screw-type positive displacement pump (see Chapter 6).

To achieve a seal between the rotors, oil is injected into the compressor: to prevent this being carried into the system, the oil separator is larger and more complex than the normal delivery oil separator associated with a reciprocating compressor. Also, because some of the heat of compression is transferred to the oil, a larger oil cooler has to be fitted, which may be either water or refrigerant cooled.

Since a.c. motor driven compressors are usually single speed, some form of cylinder unloading gear is necessary to reduce the compressor capacity. This unloading gear usually comprises a means of holding the suction valves open.

Condensers

Condensers are generally water cooled, as mentioned previously, and are of the shell and tube type. A typical modern unit is shown in Figure 9.4 in which it will be seen that the refrigerant passes over the tubes and the cooling water is passed through the tubes. In the case of sea water cooled condensers it is usual to have a two-pass arrangement through the tubes. The sea water side maintenance mentioned for coolers in Chapter 7 applies also to this condenser.

Where condensers are of 3 m and over in length between tube plates it is quite usual to have a double refrigerant liquid outlet so that the refrigerant drains away easily when the vessel is pitching or rolling.

Evaporators

Evaporators fall into two categories: refrigerant to air and refrigerant to secondary refrigerant types.

The most simple of the refrigerant to air type is in the form of a bank of tubes with an extended surface of gills or fins. In these the refrigerant is expanded in the tubes while the air is passed over the fins by circulating fans. This type of unit will be found in the domestic cold stores in which the fan and coil unit are one, and a larger version in direct expansion cargo or air conditioning systems where the fan or fans may be remote.

A more elaborate design is used for secondary refrigerant cooling which takes the form of a shell and tube vessel. Such a type is illustrated in Figure 9.5 and employs direct expansion. In this case the refrigerant

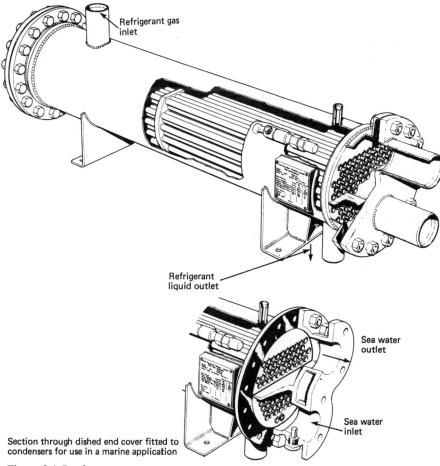

Refrigerant gas inlet

Refrigerant liquid outlet

Sea water outlet

Sea water inlet

Section through dished end cover fitted to condensers for use in a marine application

Figure 9.4 Condenser

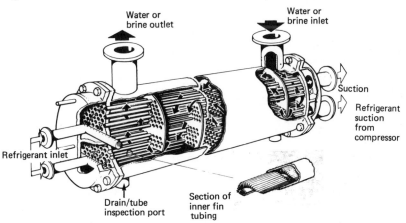

Water or brine outlet

Water or brine inlet

Suction

Refrigerant suction from compressor

Refrigerant inlet

Drain/tube inspection port

Section of inner fin tubing

Figure 9.5 Evaporator

passes through the tubes and the secondary refrigerant is passed over
the tube bank. The refrigerant is sprayed into the tubes so as to ensure
an even distribution through all the tubes. Any oil present is not sprayed
and drains away. In this type of evaporator two features are employed to
improve heat transfer efficiency. On the refrigerant side there is a
centre tube with a spiral fin fitted around it (as illustrated) or the insert
may be in the form of an aluminium star which has a spiral twist on it.
Also, baffles are arranged on the brine side to deflect the brine across
the tube bank.

Refrigerant flow control valves

It is usual to have a solenoid valve in the liquid line prior to the
expansion valve or regulator. This shuts or opens as determined by the
thermostat in the space or the secondary refrigerant being cooled. It
may also be used to shut off various circuits in a cooler when the
machine is operating on part-load conditions.

The expansion valve/regulator is a more complex piece of equipment
which meters the flow of refrigerant from the high-pressure to the
low-pressure side of the system. This may be of the thermostatic type, as
shown in Figure 9.6. The bulb senses the temperature of the refrigerant
at the outlet from the evaporator and opens or closes the valve
accordingly. The design of the valve is critical and is related to the
pressure difference between the delivery and expansion side. Therefore, it is essential that the delivery pressure is maintained at or near the

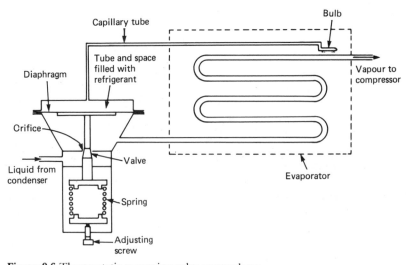

Figure 9.6 Thermostatic expansion valve or regulator

maximum design pressure. Thus, if the vessel is operating in cold sea water temperatures it is necessary to re-circulate the cooling water to maintain the correct delivery pressure from the condenser. If this is not done, the valve will 'hunt' and refrigerant liquid may be returned to the compressor suction.

Ancillary fittings

Delivery oil separators are essential for screw compressors, but for other systems, depending on the design criteria and length of pipe run, they may or may not be fitted.

Refrigerant driers are essential with the Freon gases to remove water from the system, otherwise freezing of the water can take place in the expansion valve.

A liquid receiver may be fitted for two reasons. Firstly, to give a sufficient reserve of refrigerant in the system to cater for various operating conditions (this is known as a back-up receiver). Secondly, for storage of the refrigerant where it is required to pump over, i.e. store, the charge for maintenance purposes. In very small systems this pump over can sometimes be achieved in the condenser.

Cargo refrigeration

Refrigerated cargo vessels usually require a system which provides for various spaces to be cooled to different temperatures. The arrangements adopted can be considered in three parts: the central primary refrigerating plant, the brine circulating system, and the air circulating system for cooling the cargo in the hold.

A central refrigerating plant is shown in Figure 9.7. The refrigerant flow through the chiller splits into four circuits, each with its own expansion valve. The four circuits are used to control the amount of evaporator surface, depending on the degree of condenser loading at the time, thus giving greater system flexibility. The large oil separator is a feature of screw compressor plants and the circuit for oil return is shown in the illustration.

Each primary refrigerant circuit has its own evaporator within the brine chiller (as shown in Figure 9.7) which results in totally independent gas systems. There will probably be three such systems on a cargo or container ship installation. Since they are totally independent each system can be set to control the outlet brine at different temperatures. Each brine temperature is identified by a colour and will have its own circulating pump. The cold brine is supplied to the cargo space air cooler and the flow of this brine is controlled by the temperature of the air leaving the cooler.

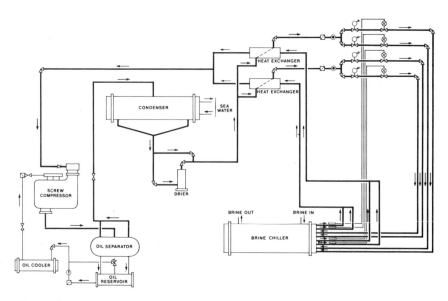

Figure 9.7 Central refrigerating plant

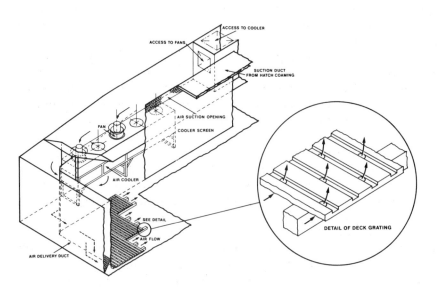

Figure 9.8 Cargo space arrangements

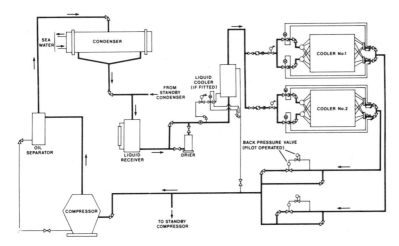

Figure 9.9 Direct expansion system

The cooler in the cargo space is arranged for air circulation over it and then through the cargo before returning. An arrangement of fans and ducting direct the air to the cooler and below the cargo (Figure 9.8). The cargo is stacked on gratings which allow the passage of cooled air up through the cargo.

For small refrigerated cargo spaces or provision rooms a direct expansion primary refrigerant system may be used (Figure 9.9). The twin circuit arrangement for each cooler (evaporator) provides flexibility and duplication in the event of one system failing. The back pressure valve maintains a minimum constant pressure or temperature in the evaporator when working a space in high-temperature conditions to prevent under-cooling of the cargo. If one space is operating at a low-temperature condition at the same time the back pressure valve would be bypassed. The liquid cooler illustrated in the diagram is necessary where an abnormal high static head has to be overcome between the machinery and the coolers. In this vessel the liquid is sub-cooled to prevent it flashing off before reaching the thermostatic expansion valve.

Containers which require refrigeration present particular problems. Where only a few are carried or the ship has no built-in arrangement for refrigerating containers, then clip-on or integral refrigeration plants would be provided. The clip-on or integral unit may be either air or water cooled. In the case of air cooled units adequate ventilation has to be supplied if they are fitted below decks. For water cooled units some sort of cooling water arrangement must be coupled up to each unit. Also an electrical supply is required for each type.

Vessels designed for specific refrigerated container trades have built-in ducting systems. These can be in two forms: a horizontal finger duct system in which up to 48 containers are fed from one cooler situated in the wings of the ship or, alternatively, a vertical duct system in which each stack of containers has its own duct and cooler. This type of system is employed for standard containers having two port holes in the wall opposite the loading doors. Air is delivered into the bottom opening and, after passing through a plenum, rises through a floor grating over the cargo and returns via another section of the plenum to the top port. The connection between the duct and containers is made by couplings which are pneumatically controlled.

System faults

During operation a number of particular problems can occur which will affect the plant performance.

An overcharge or excess refrigerant in the system will be seen as a high condenser pressure. The refrigerant should be pumped to the condenser and the excess released from there.

Air in the system will also show as a high condenser pressure. With the condenser liquid outlet closed the refrigerant charge should be pumped in and cooled. Releasing the purge valve will vent off the air which will have collected above the refrigerant.

Under-charge will show as a low compressor pressure and large bubbles in the liquid line sightglass. A leak test should then be carried out over the system to determine the fault and enable its rectification. A leak detector lamp for Freon refrigerants may be of the methylated spirit type, but more commonly uses Calor gas (butane/propane). The Freon is drawn into the flame and the flame will change colour, going from green to blue depending on the concentration of the gas.

When charging the system with more gas the main liquid valve should be closed and gas introduced before the regulating valve until the system is correctly charged. (It is possible to charge on the outlet side of the regulating valve and is quicker, but this requires a good amount of experience to prevent liquid carrying over and damaging the compressor.)

Moisture in the system may change to ice and close up the regulating valve, resulting in a drop in pressure on the evaporator side and a rise in pressure on the condenser side. The drier should be examined and the drying chemicals will probably require replacing. A correctly operating regulating valve will have frosting on the outlet side but not on the inlet side.

Air conditioning

Ships travel the world and are therefore subject to various climatic conditions. The crew of the ship must be provided with reasonable conditions in which to work regardless of the weather. Temperature alone is not a sufficient measure of conditions acceptable to the human body. Relative humidity in conjunction with temperature more truly determines the environment for human comfort. Relative humidity, expressed as a percentage, is the ratio of the water vapour pressure in the air tested, to the saturated vapour pressure of air at the same temperature. The fact that less water can be absorbed as air is cooled and more can be absorbed when it is heated is the major consideration in air conditioning system design. Other factors are the nearness of heat sources, exposure to sunlight, sources of cold and the insulation provided around the space.

An air conditioning system aims to provide a comfortable working environment regardless of outside conditions. Satisfactory air treatment must involve a relatively 'closed' system where the air is circulated and returned. However, some air is 'consumed' by humans and some machinery so there is a requirement for renewal. Public rooms and accommodation will operate with a reduced percentage of air renewal since the conditioning cost of 100% renewal would be considerable. Galleys and sanitary spaces, for instance, must have 100% renewal, but here the air quantities and treatment costs will be much smaller. Systems may however be designed for 100% renewal of air although not necessarily operated in this way. Noise and vibration from equipment used in the system should be kept to a minimum to avoid a different kind of discomfort. Three main types of marine air conditioning system are in general use, the *single duct*, the *twin duct* and the *single duct with reheat*.

The single-duct system is widely used on cargo ships (Figure 9.10). Several central units are used to distribute conditioned air to a number of cabins or spaces via a single pipe or duct. In warm climates a mixture of fresh and recirulated air is cooled and dehumidified (some water is removed) during its passage over the refrigeration unit. In cold climates the air mixture is warmed and humidified either by steam, hot water or electric heating elements. The temperature and humidity of the air is controlled automatically at the central unit. Within the conditioned space control is by variation of the volume flow of air.

The twin-duct system provides increased flexibility and is mainly used on passenger ships (Figure 9.11). A central unit is used with cooled dehumidified air provided through one duct. The other duct is supplied with cooled air that has been reheated. Each treated space is provided with a supply from each duct which may be mixed as required at the

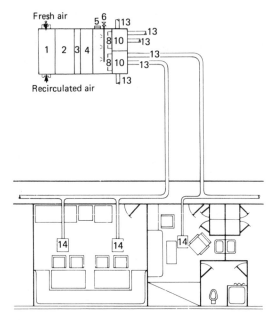

Figure 9.10 Single-duct system

1 Mixing box
2 Fan
3 Filter
4 Cooler
5 Pressure relief valve
6 Humidifier
7 Pre-heater
8 Zone heaters
9 Re-heater
10 Plenums
11 Warmer air plenums
12 Cooler air plenums
13 Pre-insulated spiro ducting
14 Air terminals
15 Air terminals with mixture
    control

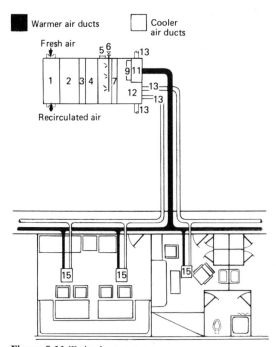

Figure 9.11 Twin-duct system

outlet terminal. In cold climates the preheater will warm both supplies of air, resulting in a warm and a hot supply to each space.

The 'single duct with reheat' system is used for vessels operating in mainly cool climates. The central unit will cool and dehumidify or preheat and humidify the air as required by outside conditions. In addition, before discharge into the treated space a local reheating unit will heat the air if required, depending upon the room thermostat setting.

The refrigeration system used in the central unit is shown in Figure 9.12. A direct-expansion system is shown using a reciprocating

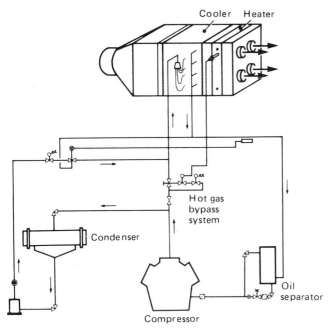

Figure 9.12 Direct-expansion refrigeration system for an air cooler

compressor, sea water cooled condenser and a thermostatically controlled regulating valve. The air to be cooled passes over the evaporator or cooler. The cooling effect of the unit may need to be reduced if there is no great demand and the hot gas bypass system provides this facility.

Maintenance of the above systems will involve the usual checks on the running machinery and the cleaning of filters. Air filters in the central units are usually washable but may be disposable. The filters should be attended to as required, depending upon the location of the ship.

Ventilation

Ventilation is the provision of a supply of fresh untreated air through a space. Natural ventilation occurs when changes in temperature or air density cause circulation in the space. Mechanical or forced ventilation uses fans for a positive movement of large quantities of air.

Natural ventilation is used for some small workshops and stores but is impractical for working areas where machinery is present or a number of people are employed.

Forced ventilation may be used in cargo spaces where the movement of air removes moisture or avoids condensation, removes odours or gases, etc.

The machinery space presents another area which requires ventilation. As a result of its large size and the fact that large volumes of air are consumed a treatment plant would be extremely costly to run. Ventilation is therefore provided in sufficient quantities for machinery air consumption and also to effect cooling. The usual distribution arrangement is shown in Figure 9.13.

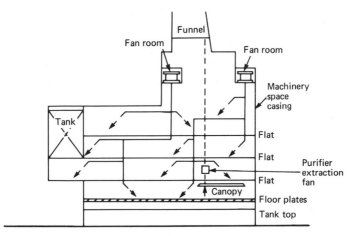

Figure 9.13 Machinery space ventilation—diagrammatic

Several axial-flow fans provide air through ducting to the various working platforms. The hot air rises in the centre and leaves through louvres or openings, usually in the funnel. The machinery control room, as a separate space, may well be arranged for air conditioning with an individual unit which draws air through trunking from the outside and exhausts back to the atmosphere.

Chapter 10

Deck machinery and hull equipment

The various items of machinery and equipment found outside of the machinery space will now be described. These include deck machinery such as mooring equipment, anchor handling equipment, cargo handling equipment and hatch covers. Other items include lifeboats and liferafts, emergency equipment, watertight doors, stabilisers and bow thrusters.

The operations of mooring, cargo handling and anchor handling all involve controlled pulls or lifts using chain cables, wire or hemp ropes. The drive force and control arrangements adopted will influence the operations. Several methods are currently in use, and these will be examined before considering the associated equipment.

Three forms of power are currently in use: *steam, hydraulic* and *electric*. Each will be described in turn, together with its advantages and disadvantages for particular duties or locations.

Steam

With a steam powering and control system the steam pipelines are run along the deck to the various machines. Steam is admitted first to a directional valve and then to the steam admission valve. Double-acting steam engines, usually with two cylinders, are used to drive the machinery. Additional back pressure valves are used with mooring winches to control tension when the machine is stalled or brought to a stop by the load. Arrangements must also be made, often associated with the back pressure valve, to counteract the fluctuations in main steam line pressure as a result of other users of steam.

The steam-powered system was widely used on tankers since it presented no fire or explosion risk, but the lengths of deck pipework and the steam engines themselves presented considerable maintenance tasks which have generally resulted in their replacement by hydraulically powered equipment.

Hydraulic systems

The open-loop circuit takes oil from the tank and pumps it into the hydraulic motor. A control valve is positioned in parallel with the motor. When it is open the motor is stationary; when it is throttled or closed the motor will operate. The exhaust oil returns to the tank. This method can provide stepless control, i.e. smooth changes in motor speed.

The live-line circuit, on the contrary, maintains a high pressure from which the control valve draws pressurised oil to the hydraulic motor (in series with it), as and when required.

In the closed-loop circuit the exhaust oil is returned direct to the pump suction. Since the oil does not enter an open tank, the system is considered closed.

Low-pressure systems use the open-loop circuit and are simple in design as well as reliable. The equipment is, however, large, inefficient in operation and overheats after prolonged use.

Medium-pressure systems are favoured for marine applications, using either the open or closed circuit. Smaller installations are of the open-loop type. Where considerable amounts of hydraulic machinery are fitted the live-circuit, supplied by a centralised hydraulic power system, would be most economical.

Electrical operation

Early installations used d.c. supply with resistances in series to provide speed control (see Chapter 14). This inefficient power-wasting method was one possibility with d.c., but a better method was the use of Ward Leonard control. The high cost of all the equipment involved in Ward Leonard control and its maintenance is, however, a considerable disadvantage.

Machines operated on an a.c. supply require a means of speed control with either pole-changing or slip-ring motors being used. Slip-ring motors require low starting currents but waste power at less than full speed and require regular maintenance. Pole-changing motors are of squirrel cage construction, providing for perhaps three different speeds. They require large starting currents, although maintenance is negligible (see Chapter 14).

Apart from the advantages and disadvantages for each of the drive and control methods, all electric drives have difficulty with heavy continuous overloads. Each system has its advocates and careful design and choice of associated equipment can provide a satisfactory installation.

Mooring equipment

Winches with various arrangements of barrels are the usual mooring equipment used on board ships. A mooring winch is shown in Figure 10.1 where the various parts can be identified. The winch barrel or drum is used for hauling in or letting out the wires or ropes which will fasten the ship to the shore. The warp end is used when moving the ship using ropes or wires fastened to bollards ashore and wrapped around the warp end of the winch.

The construction of a mooring winch will now be examined, again with respect to Figure 10.1. The motor drive is passed through a spur

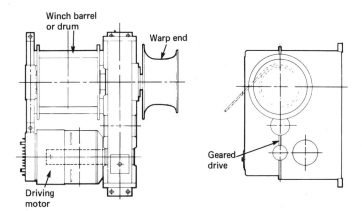

Figure 10.1 Mooring winch

gear transmission, a clutch and thus to the drum and warp end. A substantial frame supports the assembly and a band brake is used to hold the drum when required. The control arrangements for the drive motor permit forward or reverse rotation together with a selection of speeds during operation.

Modern mooring winches are arranged as automatic self-tensioning units. The flow of the tides or changes in draught due to cargo operations may result in tensioning or slackening of mooring wires. To avoid constant attention to the mooring wires the automatic self-tensioning arrangement provides for paying out (releasing) or recovering wire when a pre-set tension is not present.

Anchor handling equipment

The windlass is the usual anchor handling device where one machine may be used to handle both anchors. A more recent development,

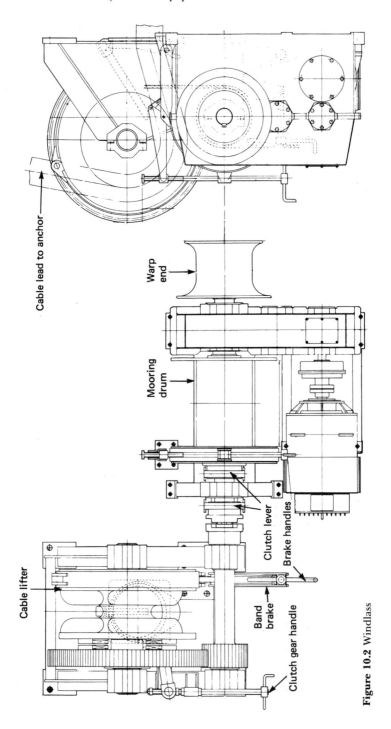

Figure 10.2 Windlass

Cable lead to anchor

Warp end

Mooring drum

Clutch lever

Brake handles

Cable lifter

Band brake

Clutch gear handle

particularly on larger vessels, is the split windlass where one machine is used for each anchor.

One unit of a split windlass is shown in Figure 10.2. The rotating units consist of a cable lifter with shaped snugs to grip the anchor cable, a mooring drum for paying out or letting go of mooring wires and a warp end for warping duties. Each of these units may be separately engaged or disengaged by means of a dog clutch, although the warp end is often driven in association with the mooring drum. A spur gear assembly transmits the motor drive to the shaft where the various dog clutches enable the power take-off. Separate band brakes are fitted to hold the cable lifter and the mooring drum when the power is switched off.

The cable lifter unit, shown in Figure 10.2, is mounted so as to raise and lower the cable from the spurling pipe, which is at the top and centre of the chain or cable locker. Details of the snugs used to grip the cable and of the band brake can be seen.

Anchor capstans are used in some installations where the cable lifter rotates about a vertical axis. Only the cable lifter unit is located on deck, the driving machinery being on the deck below. A warping end or barrel may be driven by the same unit and is positioned near the cable lifter.

Cargo handling equipment

Cargo winches are used with the various derrick systems arranged for cargo handling. The unit is rated according to the safe working load to be lifted and usually has a double-speed provision when working at half load.

In the cargo winch, spur reduction gearing transfers the motor drive to the barrel shaft. A warp end may be fitted for operating the derrick topping lift (the wire which adjusts the derrick height). Manually operated band brakes may be fitted and the drive motor will have a brake arranged to fail-safe, i.e. it will hold the load if power fails or the machine is stopped.

A derrick rig, known as 'union purchase', is shown in Figure 10.3. One derrick is positioned over the quayside and the other almost vertically over the hold. Topping wires fix the height of the derricks and stays to the deck may be used to prevent fore and aft movement. Cargo handling wires run from two winches and join at the hook. A combination of movements from the two winches enables lifting, transfer and lowering of the cargo. This is only one of several possible derrick arrangements or rigs. Although being very popular for many years, it requires considerable crew time to set up and considerable manpower for operation.

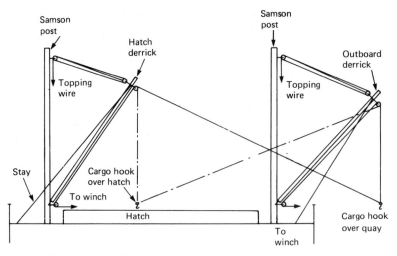

Figure 10.3 Union purchase rig

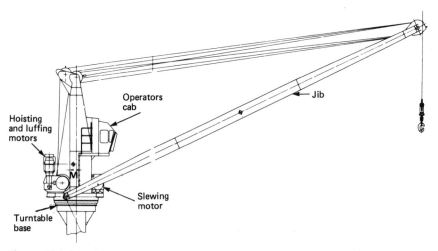

Figure 10.4 General cargo crane

Cranes have replaced derricks on many modern ships. Positioned between the holds, often on a platform which can be rotated through 360°, the deck crane provides an immediately operational unit requiring only one man to operate it. Double gearing is a feature of most designs, providing a higher speed at lighter loads. Various types of crane exist for particular duties, for example a general duties crane using a hook and a grabbing crane for use with bulk cargoes.

A general cargo crane is shown in Figure 10.4. Three separate drives provide the principal movements: a hoisting motor for lifting the load, a luffing motor for raising or lowering the jib, and a slewing motor for rotating the crane. The operator's cab is designed to provide clear views of all the cargo working area so that the crane operator can function alone. The crane is usually mounted on a pedestal to offer adequate visibility to the operator. For occasional heavy loads arrangements for two cranes to work together, i.e. twinning, can be made with a single operator using a master and slave control system in the two cranes. A common revolving platform will be necessary for this arrangement. The operating medium for deck crane motors may be hydraulic or electric, utilising circuits referred to earlier.

Maintenance

All deck machinery is exposed to the most severe aspects of the elements. Total enclosure of all working parts is usual with splash lubrication for gearing. The various bearings on the shafts will be greased by pressure grease points. Open gears and clutches are lubricated with open gear compound. Particular maintenance tasks will be associated with the type of motor drive employed.

Hatch covers

Hatch covers are used to close off the hatch opening and make it watertight. Wooden hatch covers, consisting of beams and boards over the opening and covered with tarpaulins, were once used but are no longer fitted. Steel hatch covers, comprising a number of linked steel covers, are now fitted universally. Various designs exist for particular applications, but most offer simple and quick opening and closing, which speed up the cargo handling operation.

A MacGregor single-pull weather-deck hatch cover is shown in Figure 10.5. The hatch covers are arranged to move on rollers along a track on top of the hatch coaming. The individual covers are linked together by chains and ride up and tip onto a stowage rack at the hatch end. A hydraulic power unit, operated from a control box at the hatch end, is used to open and close the hatch cover. It is possible to open and close the covers with a single wire pull from a crane or winch. Watertightness of the closed covers is achieved by pulling them down on to a compressible jointing strip. This is done by the use of cleats which may be hand-operated or automatically engaged as the hatch closes.

Hatch covers below the weather decks are arranged flush with the deck, as shown in Figure 10.6. In the arrangement shown a self-contained hydraulic power pack with reservoir pump and motor is mounted into a pair of hatch covers. This power pack serves the

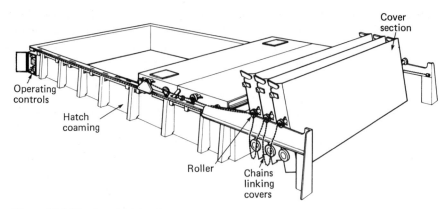

Figure 10.5 Weather-deck hatch cover

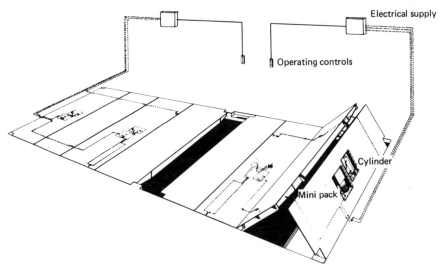

Figure 10.6 Tween-deck hatch cover

operating cylinder for the pair of covers. Control is from a nearby point and hydraulic piping is reduced to a minimum.

Maintenance requirements for this equipment are usually minimal but regular inspection and servicing should be undertaken. Most hatch covers can, if necessary, be removed manually.

Stabilising systems

There are two basic stabilising systems used on ships—the fin and the tank. A stabilising system is fitted to a ship in order to reduce the rolling motion. This is achieved by providing an opposite force to that attempting to roll the ship.

Fin stabiliser

One or more pairs of fins are fitted on a ship, one on each side, see Figure 10.7. The size or area of the fins is governed by ship factors such as breadth, draught, displacement, and so on, but is very small compared with the size of the ship. The fins may be retractable, i.e. pivoting or sliding within the ship's form, or fixed. They act to apply a righting moment to the ship as it is inclined by a wave or force on one side. The angle of tilt of the fin and the resulting moment on the ship is determined by a sensing control system. The forward speed of the ship enables the fins to generate the thrust which results in the righting moment.

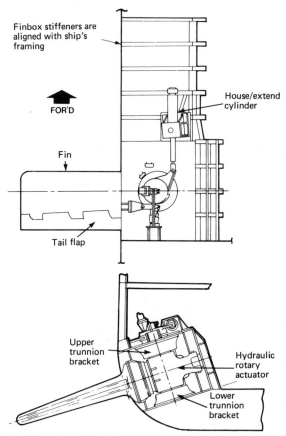

Figure 10.7 Fin stabiliser

The operating system can be compared to that of the steering gear in that a signal from the control unit causes a movement of the fin which, when it reaches the desired value, is brought to rest. The fin movement takes place as a result of a hydraulic power unit incorporating a type of variable displacement pump.

The effectiveness of the fins as stabilisers depends upon their speed of movement, which must be rapid from one extreme point to the other. The fins are rectangular in shape and streamlined in section. The use of a movable flap or a fixed and movable portion is to provide a greater restoring moment to the ship for a slightly more complicated mechanism.

The control system which signals the movement of the fins utilises two gyroscopes, one which senses movements from the vertical and the other the rolling velocity. As a result of this control system, fin movement is a function of roll angle, roll velocity, roll acceleration and natural list.

Fin stabilisers provide accurate and effective roll stabilisation in return for a complex installation, which in merchant vessels is usually limited to passenger ships. It is to be noted that at low ship speeds the stabilising power falls off, and when stationary no stabilisation is possible.

Tank stabiliser

A tank stabiliser provides a righting or anti-rolling force as a result of the delayed flow of fluid in a suitably positioned transverse tank. The system operation is independent of ship speed and will work when the ship is at rest.

Consider a mass of water in an athwartships tank. As the ship rolls the water will be moved, but a moment or two after the ship. Thus when the ship is finishing its roll and about to turn, the still moving water will oppose the return roll. The water mass thus acts against the roll at each ship movement. This athwartships tank is sometimes referred to as a 'flume'. The system is considered passive, since the water flow is activated by gravity.

A wing tank system arranged for controlled passive operation is shown in Figure 10.8. The greater height of tank at the sides permits a larger water build-up and thus a greater moment to resist the roll. The rising fluid level must not however fill the wing tank. The air duct between the two wing tanks contains valves which are operated by a roll sensing device. The differential air pressure between tanks is regulated to allow the fluid flow to be controlled and 'phased' for maximum roll stabilisation.

A tank system must be specifically designed for a particular ship by using data from model tests. The water level in the system is critical and

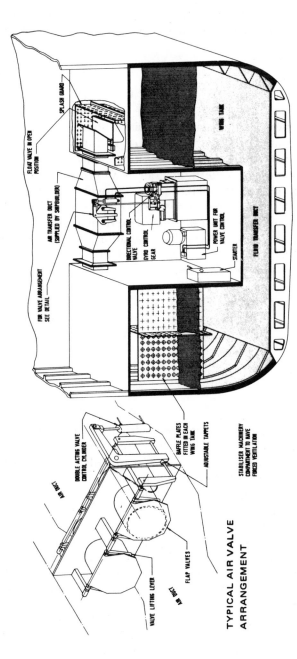

Figure 10.8 Air-controlled tank stabiliser

must be adjusted according to the ship's loaded condition. Also there is a free surface effect resulting from the moving water which effectively reduces the stability of the ship. The tank system does, however, stabilise at zero speed and is a much less complex installation than a fin stabiliser.

Watertight doors

Watertight doors are provided where an opening in a watertight bulkhead is essential. On cargo ships with a shaft tunnel, the entrance would have a watertight door fitted. All doors fitted below the waterline must be of the sliding type, arranged horizontally or vertically.

A horizontal sliding watertight door is shown in Figure 10.9. The robust frame fits into the bulkhead and provides the trackway along which the door slides. The door is moved by a hydraulic cylinder which may be power or manually operated. The door must be arranged for local opening and closing as well as operation from a point above the bulkhead deck. The power unit situated above the bulkhead deck provides either powered or hand operation of the door.

Watertight doors should be tested for operation by closing and opening during fire drill. The hydraulic system should be occasionally checked for leaks and to ensure sufficient oil is present in the system. The bottom trackway of the door should be checked for cleanliness and freedom from obstructing matter.

Bow thruster

The bow thruster is a propulsion device fitted to certain types of ships to improve manoeuvrability. The thrust unit consists of a propeller mounted in an athwartships tunnel and provided with some auxiliary drive such as an electric or hydraulic motor. During operation water is forced through the tunnel to push the ship sideways either to port or starboard as required. The unit is normally bridge controlled and is most effective when the vessel is stationary.

A controllable-pitch type thruster unit is shown in Figure 10.10. A servo motor located in the gear housing enables the propeller blade pitch to be altered, to provide water flow in either direction. With this arrangement any non-reversing prime mover, like a single-speed electric motor, may be used. The prime mover need not be stopped during manoeuvring operations since the blades can be placed at zero pitch when no thrust is desired. The drive is obtained through a flexible drive shaft, couplings and bevel gears. Special seals prevent any sea water leakage into the unit. The complete assembly includes part of the

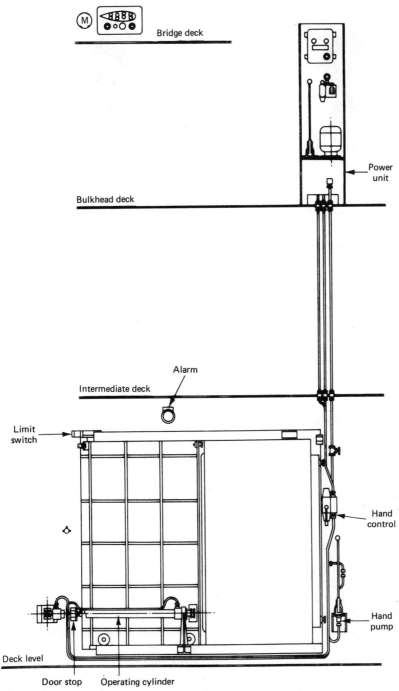

Figure 10.9 Horizontal sliding watertight door

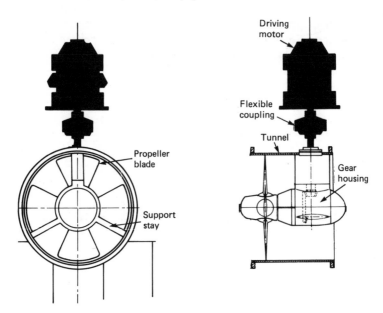

Figure 10.10 Bow thruster

athwartships tunnel through which water is directed to provide the thrust.

Safety equipment

A number of items will now be considered under this general heading. They are emergency equipment for power generation and pumping, survival equipment such as lifeboats and liferafts, and the sound signal equipment in the form of the whistle.

Emergency equipment

Emergency equipment is arranged to operate independently of all main power sources. It includes such items as the emergency generator and the emergency fire pump. Both items of machinery are located remote from the engine room and usually above the bulkhead deck, that is at the weather deck level or above. The emergency generator is usually on one of the accommodation decks while the emergency fire pump is often inside the forecastle.

The emergency generator is a diesel-driven generator of sufficient capacity to provide essential circuits such as steering, navigation lights and communications. The diesel engine has its own supply system, usually of light diesel oil for easy starting. Batteries, compressed air or an hydraulic accumulator may be used for starting the machine. Small machines may be air cooled but larger units are arranged usually for water cooling with an air cooled radiator as heat exchanger in the system. A small switchboard is located in the same compartment to connect the supply to the various emergency services (see Chapter 14).

Modern systems are arranged to start up the emergency generator automatically when the main power supply fails. The system should be checked regularly and operated to ensure its availability if required. Fuel tanks should be kept full, ample cooling water should be in the radiator cooling system, and the starting equipment should be functional. Batteries of course, should be fully charged or air receivers full.

The emergency fire pump is arranged to supply the ship's fire main when the machinery space pump is not available. One possible arrangement, as used on large tankers, is shown diagrammatically in Figure 10.11. A diesel engine with its own fuel supply system, starting

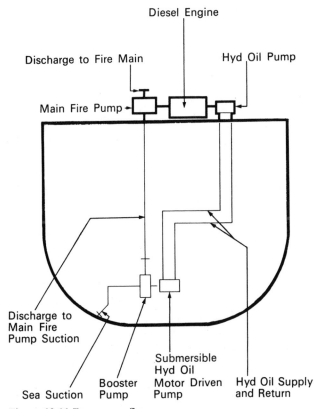

Figure 10.11 Emergency fire pump

arrangements, etc., drives at one end a main fire pump and at the other an hydraulic oil pump. The hydraulic oil pump supplies a hydraulic motor, located low down in the ship, which in turn operates a sea water booster pump. The booster pump has its own sea suction and discharges to the main pump suction. The main pump then supplies sea water to the fire main. The booster pump arrangement is necessary because of the considerable depth of many large modern ships.

Survival equipment

Lifesaving equipment on board ship, apart from smaller items such as lifebuoys and lifejackets, consists of lifeboats and liferafts. Lifeboats are rigid vessels secured into davits which enable the boat to be launched over the ship's side. Liferafts are inflatable vessels, usually stowed on deck in canisters which must be thrown overboard, whereupon they are automatically inflated.

Lifeboat accommodation for all the ship's crew must be provided on both sides of the ship. This is to allow for a situation when only the boats on one side can be lowered. The boats must be more than 7.3 m long and

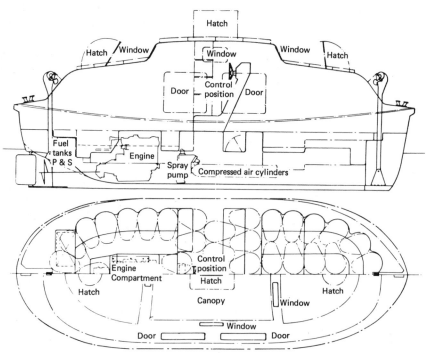

Figure 10.12 Lifeboat

carry sufficient equipment and provisions for survival for a reasonable period (Figure 10.12). This would include oars, a boat hook, a compass, distress rockets, first aid equipment, rations and fresh water. They must also be partially or totally enclosed, self-righting and equipped with an engine.

Lifeboats on cargo ships of 20 000 tons gross tonnage and above must be capable of being launched when the ship is making headway at speeds up to 5 knots.

A new requirement for all new ships is that a rescue boat, capable of being launched in five minutes, must be carried. This boat is to be used to rescue persons from the sea and also to gather together the lifeboats.

Lifeboat davits are provided as stowage for the lifeboats which can readily be released to lower the boats without any mechanical assistance. 'Gravity davits', as they are called, slide down and position the boat for lowering as soon as they are released. The davits must be able to lower the boats when the ship is heeled to 15° on either side.

One type of gravity davit is shown in Figure 10.13. The lifeboat is held against the cradle by ropes called 'gripes'. Another wire, either separate or combined with the gripes, holds the cradle in its upper position. With the gripes and the cradle securing device free, the winch handbrake can be released to enable the cradle to slide down and over the ship's side. A tricing-in pendant (a wire) bring the lifeboat close to the ship's side to enable it to be boarded. The bowsing lines which fasten to each end of the lifeboat are then used to hold it in to the ship's side, the tricing pendant then being released. Once the crew are on board the bowsing lines are released and the lifeboat lowered to the water. The wires which raise or lower the boat are called 'falls' and the speed of descent is restricted to 36 m/min by a centrifugal brake. The handbrake used to lower the boat has a 'dead man's handle' or weighted lever, which, if released, will apply the brake.

Liferafts are normally provided to accommodate all of the ship's complement. They are usually stored in cylindrical glass-reinforced plastic containers which are secured on chocks on the deck. Inflation takes place automatically when the container is thrown overboard, the container bursts open and the liferaft floats clear. A pressurised cylinder of carbon dioxide is used to inflate the raft. One type of liferaft is shown in Figure 10.14, where it can be seen to be a well equipped totally enclosed arrangement. The survival equipment located in the raft is similar to that provided in lifeboats. Liferafts must normally be boarded from the water unless they are of a special type which is lowered, fully inflated, by a davit; but it is not usual to fit this type on cargo vessels. Liferafts must be stowed in such a way that they will float free and inflate if the vessel sinks. A hydrostatic release is normally used which releases the lashings at a predetermined depth of water.

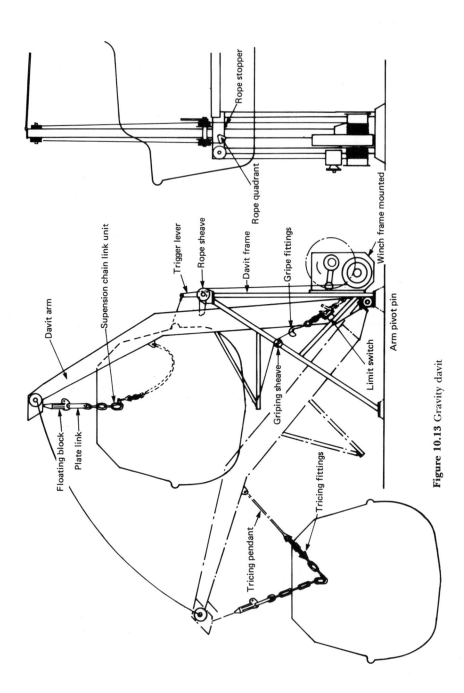

Figure 10.13 Gravity davit

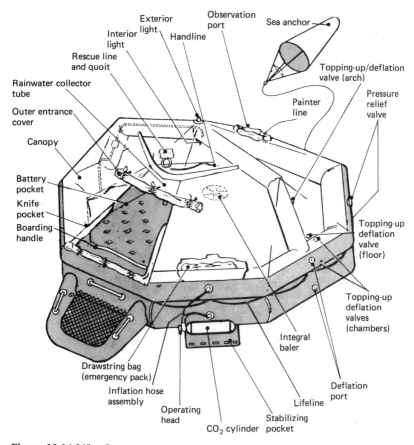

Figure 10.14 Liferaft

Whistle

International regulations require audible signals to be made by a ship in conditions of poor visibility. The ship's whistle is provided and arranged to give prolonged blasts at timed intervals when operated by a hand control.

One type of air-operated whistle is shown in Figure 10.15. The compressed air acting on the diaphragm causes it to vibrate and the sound waves are amplified in the horn. The control system associated with the whistle can provide whistle operation as long as any of the operating switches is in the 'on' position. Alternatively short blasts can be given by on-off operation, since instantaneous cut-off occurs after each blast. A more sophisticated control system incorporates timing gear which provides a prolonged blast every two minutes, or other

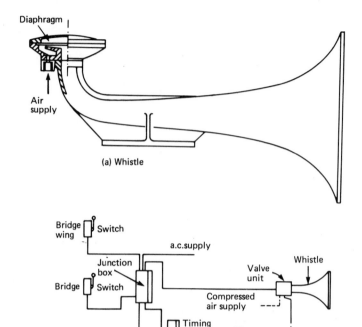

Diaphragm

Air supply

(a) Whistle

Bridge wing — Switch

a.c.supply

Junction box

Bridge — Switch

Valve unit

Whistle

Compressed air supply

Timing box

Bridge wing — Switch

Push button (bridge)

Lanyard for emergency operation

(b) Signal control system

Figure 10.15 Whistle

arrangements as required. The whistle switches are usually on the bridge wings and inside the bridge. The whistle is also arranged for direct operation from a lanyard which extends from the bridge. The compressed air supply can vary in pressure over a considerable range without affecting the whistle operation.

Chapter 11
Shafting and propellers

The transmission system on a ship transmits power from the engine to the propeller. It is made up of shafts, bearings, and finally the propeller itself. The thrust from the propeller is transferred to the ship through the transmission system.

The different items in the system include the thrust shaft, one or more intermediate shafts and the tailshaft. These shafts are supported by the thrust block, intermediate bearings and the sterntube bearing. A sealing arrangement is provided at either end of the tailshaft with the propeller and cone completing the arrangement. These parts, their location and purpose are shown in Figure 11.1.

Thrust block

The thrust block transfers the thrust from the propeller to the hull of the ship. It must therefore be solidly constructed and mounted onto a rigid seating or framework to perform its task. It may be an independent unit or an integral part of the main propulsion engine. Both ahead and astern thrusts must be catered for and the construction must be strong enough to withstand normal and shock loads.

The casing of the independent thrust block is in two halves which are joined by fitted bolts (Figure 11.2). The thrust loading is carried by bearing pads which are arranged to pivot or tilt. The pads are mounted in holders or carriers and faced with white metal. In the arrangement shown the thrust pads extend threequarters of the distance around the collar and transmit all thrust to the lower half of the casing. Other designs employ a complete ring of pads. An oil scraper deflects the oil lifted by the thrust collar and directs it onto the pad stops. From here it cascades over the thrust pads and bearings. The thrust shaft is manufactured with integral flanges for bolting to the engine or gearbox shaft and the intermediate shafting, and a thrust collar for absorbing the thrust.

199

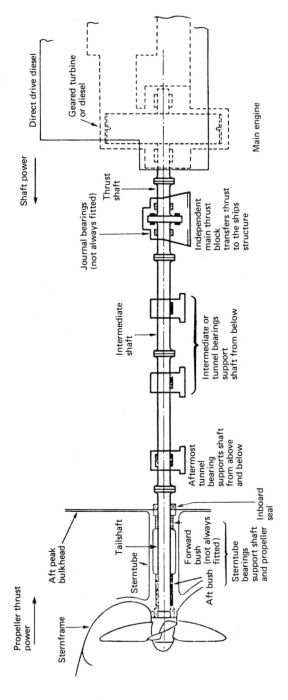

Figure 11.1 Transmission system

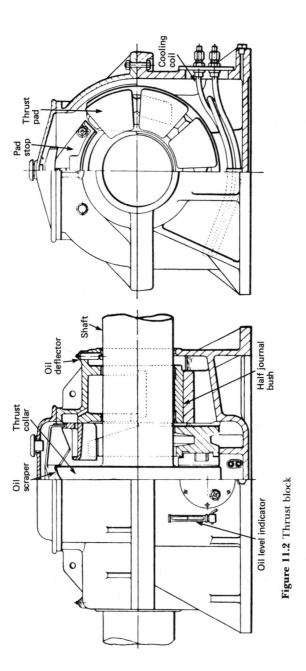

Figure 11.2 Thrust block

Where the thrust shaft is an integral part of the engine, the casing is usually fabricated in a similar manner to the engine bedplate to which it is bolted. Pressurised lubrication from the engine lubricating oil system is provided and most other details of construction are similar to the independent type of thrust block.

Shaft bearings

Shaft bearings are of two types, the aftermost tunnel bearing and all others. The aftermost tunnel bearing has a top and bottom bearing shell because it must counteract the propeller mass and take a vertical upward thrust at the forward end of the tailshaft. The other shaft bearings only support the shaft weight and thus have only lower half bearing shells.

An intermediate tunnel bearing is shown in Figure 11.3. The usual journal bush is here replaced by pivoting pads. The tilting pad is better able to carry high overloads and retain a thick oil lubrication film. Lubrication is from a bath in the lower half of the casing, and an oil thrower ring dips into the oil and carries it round the shaft as it rotates. Cooling of the bearing is by water circulating through a tube cooler in the bottom of the casing.

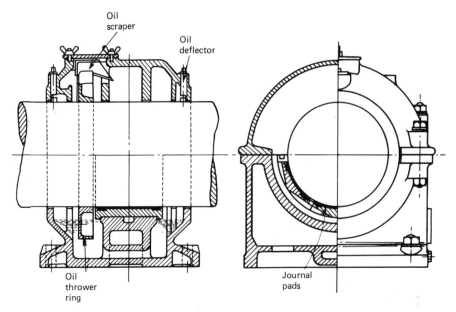

Figure 11.3 Tunnel bearing

Sterntube bearing

The sterntube bearing serves two important purposes. It supports the tailshaft and a considerable proportion of the propeller weight. It also acts as a gland to prevent the entry of sea water to the machinery space.

Early arrangements used bearing materials such as lignum vitae (a very dense form of timber) which were lubricated by sea water. Most modern designs use an oil lubrication arrangement for a white metal lined sterntube bearing. One arrangement is shown in Figure 11.4.

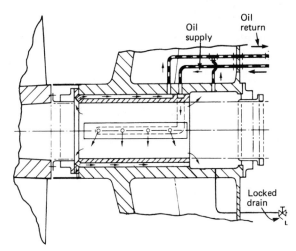

Figure 11.4 Oil lubricated sterntube bearing

Oil is pumped to the bush through external axial grooves and passes through holes on each side into internal axial passages. The oil leaves from the ends of the bush and circulates back to the pump and the cooler. One of two header tanks will provide a back pressure in the system and a period of oil supply in the event of pump failure. A low-level alarm will be fitted to each header tank.

Oil pressure in the lubrication system is higher than the static sea water head to ensure that sea water cannot enter the sterntube in the event of seal failure.

Sterntube seals

Special seals are fitted at the outboard and inboard ends of the tailshaft. They are arranged to prevent the entry of sea water and also the loss of lubricating oil from the stern bearing.

Older designs, usually associated with sea water lubricated stern bearings, made use of a conventional stuffing box and gland at the after bulkhead. Oil-lubricated stern bearings use either lip or radial face seals or a combination of the two.

Lip seals are shaped rings of material with a projecting lip or edge which is held in contact with a shaft to prevent oil leakage or water entry. A number of lip seals are usually fitted depending upon the particular application.

Face seals use a pair of mating radial faces to seal against leakage. One face is stationary and the other rotates. The rotating face of the after seal is usually secured to the propeller boss. The stationary face of the forward or inboard seal is the after bulkhead. A spring arrangement forces the stationary and rotating faces together.

Shafting

There may be one or more sections of intermediate shafting between the thrust shaft and the tailshaft, depending upon the machinery space location. All shafting is manufactured from solid forged ingot steel with integral flanged couplings. The shafting sections are joined by solid forged steel fitted bolts.

The intermediate shafting has flanges at each end and may be increased in diameter where it is supported by bearings.

The propeller shaft or tailshaft has a flanged face where it joins the intermediate shafting. The other end is tapered to suit a similar taper on the propeller boss. The tapered end will also be threaded to take a nut which holds the propeller in place.

Propeller

The propeller consists of a boss with several blades of helicoidal form attached to it. When rotated it 'screws' or thrusts its way through the water by giving momentum to the column of water passing through it. The thrust is transmitted along the shafting to the thrust block and finally to the ship's structure.

A solid fixed-pitch propeller is shown in Figure 11.5. Although usually described as fixed, the pitch does vary with increasing radius from the boss. The pitch at any point is fixed, however, and for calculation purposes a mean or average value is used.

A propeller which turns clockwise when viewed from aft is considered right-handed and most single-screw ships have right-handed propellers. A twin-screw ship will usually have a right-handed starboard propeller and a left-handed port propeller.

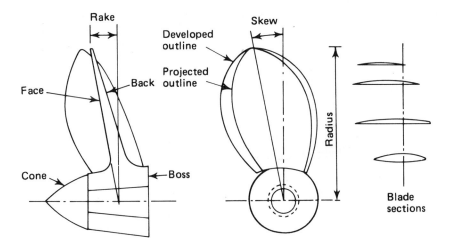

Figure 11.5 Solid propeller

Propeller mounting

The propeller is fitted onto a taper on the tailshaft and a key may be inserted between the two: alternatively a keyless arrangement may be used. A large nut is fastened and locked in place on the end of the tailshaft: a cone is then bolted over the end of the tailshaft to provide a smooth flow of water from the propeller.

One method of keyless propeller fitting is the oil injection system. The propeller bore has a series of axial and circumferential grooves machined into it. High-pressure oil is injected between the tapered section of the tailshaft and the propeller. This reduces the friction between the two parts and the propeller is pushed up the shaft taper by a hydraulic jacking ring. Once the propeller is positioned the oil pressure is released and the oil runs back, leaving the shaft and propeller securely fastened together.

The Pilgrim Nut is a patented device which provides a predetermined frictional grip between the propeller and its shaft. With this arrangement the engine torque may be transmitted without loading the key, where it is fitted. The Pilgrim Nut is, in effect, a threaded hydraulic jack which is screwed onto the tailshaft (Figure 11.6). A steel ring receives thrust from a hydraulically pressurised nitrile rubber tyre. This thrust is applied to the propeller to force it onto the tapered tailshaft. Propeller removal is achieved by reversing the Pilgrim Nut and using a withdrawal plate which is fastened to the propeller boss by studs. When

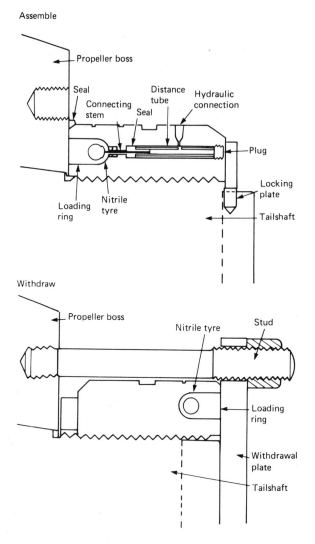

Figure 11.6 Pilgrim Nut operation

the tyre is pressurised the propeller is drawn off the taper. Assembly
and withdrawal are shown in Figure 11.6.

Controllable-pitch propeller

A controllable-pitch propeller is made up of a boss with separate blades
mounted into it. An internal mechanism enables the blades to be moved

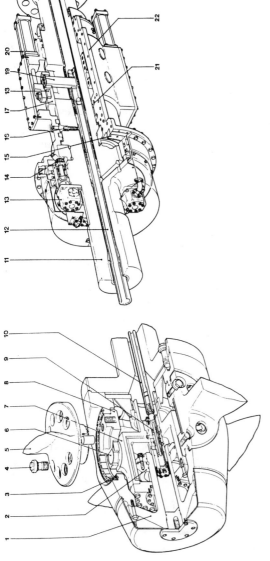

Figure 11.7 Controllable-pitch propeller

1 Piston rod
2 Piston
3 Blade seal
4 Blade bolt
5 Blade
6 Crank pin
7 Servo motor cylinder
8 Crank ring
9 Control valve
10 Valve rod
11 Mainshaft

12 Valve rod
13 Main pump
14 Pinion
15 Internally toothed gear ring
16 Non-return valve
17 Sliding ring
18 Sliding thrust block
19 Corner pin
20 Auxiliary servo motor
21 Pressure seal
22 Casing

simultaneously through an arc to change the pitch angle and therefore the pitch. A typical arrangement is shown in Figure 11.7.

When a pitch demand signal is received a spool valve is operated which controls the supply of low-pressure oil to the auxiliary servo motor. The auxiliary servo motor moves the sliding thrust block assembly to position the valve rod which extends into the propeller hub. The valve rod admits high-pressure oil into one side or the other of the main servo motor cylinder. The cylinder movement is transferred by a crank pin and ring to the propeller blades. The propeller blades all rotate together until the feedback signal balances the demand signal and the low-pressure oil to the auxiliary servo motor is cut off. To enable emergency control of propeller pitch in the event of loss of power the spool valves can be operated by hand. The oil pumps are shaft driven.

The control mechanism, which is usually hydraulic, passes through the tailshaft and operation is usually from the bridge. Varying the pitch will vary the thrust provided, and since a zero pitch position exists the engine shaft may turn continuously. The blades may rotate to provide astern thrust and therefore the engine does not require to be reversed.

Cavitation

Cavitation, the forming and bursting of vapour-filled cavities or bubbles, can occur as a result of pressure variations on the back of a propeller blade. The results are a loss of thrust, erosion of the blade surface, vibrations in the afterbody of the ship and noise. It is usually limited to high-speed heavily loaded propellers and is not a problem under normal operating conditions with a well designed propeller.

Propeller maintenance

When a ship is in dry dock the opportunity should be taken to thoroughly examine the propeller, and any repairs necessary should be carried out by skilled dockyard staff.

A careful examination should be made around the blade edges for signs of cracks. Even the smallest of cracks should not be ignored as they act to increase stresses locally and can result in the loss of a blade if the propeller receives a sharp blow. Edge cracks should be welded up with suitable electrodes.

Bent blades, particularly at the tips, should receive attention as soon as possible. Except for slight deformation the application of heat will be required. This must be followed by more general heating in order to stress relieve the area around the repair.

Surface roughness caused by slight pitting can be lightly ground out and the area polished. More serious damage should be made good by welding and subsequent heat treatment. A temporary repair for deep pits or holes could be done with a suitable resin filler.

Chapter 12
Steering gear

The steering gear provides a movement of the rudder in response to a signal from the bridge. The total system may be considered made up of three parts, *control equipment, a power unit* and a *transmission to the rudder stock.* The control equipment conveys a signal of desired rudder angle from the bridge and activates the power unit and transmission system until the desired angle is reached. The power unit provides the force, when required and with immediate effect, to move the rudder to the desired angle. The transmission system, the steering gear, is the means by which the movement of the rudder is accomplished.

Certain requirements must currently be met by a ship's steering system. There must be two independent means of steering, although where two identical power units are provided an auxiliary unit is not required. The power and torque capability must be such that the rudder can be swung from 35° one side to 35° the other side with the ship at maximum speed, and also the time to swing from 35° one side to 30° the other side must not exceed 28 seconds. The system must be protected from shock loading and have pipework which is exclusive to it as well as be constructed from approved materials. Control of the steering gear must be provided in the steering gear compartment.

Tankers of 10 000 ton gross tonnage and upwards must have two independent steering gear control systems which are operated from the bridge. Where one fails, changeover to the other must be immediate and achieved from the bridge position. The steering gear itself must comprise two independent systems where a failure of one results in an automatic changeover to the other within 45 seconds. Any of these failures should result in audible and visual alarms on the bridge.

Steering gears can be arranged with hydraulic control equipment known as a 'telemotor', or with electrical control equipment. The power unit may in turn be hydraulic or electrically operated. Each of these units will be considered in turn, with the hydraulic unit pump being considered first. A pump is required in the hydraulic system which can immediately pump fluid in order to provide a hydraulic force that will move the rudder. Instant response does not allow time for the pump to be switched on and therefore a constantly running pump is required which pumps fluid only when required. A variable delivery pump provides this facility.

Variable delivery pumps

A number of different designs of variable delivery pump exist. Each has a means of altering the pump stroke so that the amount of oil displaced will vary from zero to some designed maximum value. This is achieved by use of a floating ring, a swash plate or a slipper pad.

The radial cylinder (Hele-Shaw) pump is shown in. Figure 12.1. Within the casing a short length of shaft drives the cylinder body which rotates around a central valve or tube arrangement and is supported at the ends by ball bearings. The cylinder body is connected to the central valve arrangement by ports which lead to connections at the outer casing for the supply and delivery of oil. A number of pistons fit in the radial cylinders and are fastened to slippers by a gudgeon pin. The slippers fit into a track in the circular floating ring. This ring may rotate, being supported by ball bearings, and can also move from side to side since the bearings are mounted in guide blocks. Two spindles which pass out of the pump casing control the movement of the ring.

The operating principle will now be described by reference to Figure 12.2. When the circular floating ring is concentric with the central valve arrangement the pistons have no relative reciprocating motion in their cylinders (Figure 12.2(a)). As a result no oil is pumped and the pump, although rotating, is not delivering any fluid. If however the circular floating ring is pulled to the right then a relative reciprocating motion of the pistons in their cylinders does occur (Figure 12.2(b)). The lower piston, for instance, as it moves inwards will discharge fluid out through the lower port in the central valve arrangement. As it continues past the horizontal position the piston moves outwards, drawing in fluid from the upper port. Once past the horizontal position on the opposite side, it begins to discharge the fluid. If the circular floating ring were pushed to the left then the suction and discharge ports would be reversed (Figure 12.2(c)).

This pump arrangement therefore provides, for a constantly rotating unit, a no-flow condition and infinitely variable delivery in either direction. The pump is also a positive displacement unit. Where two pumps are fitted in a system and only one is operating, reverse operation might occur. Non-reversing locking gear is provided as part of the flexible coupling and is automatic in operation. When a pump is stopped the locking gear comes into action; as the pump is started the locking gear releases.

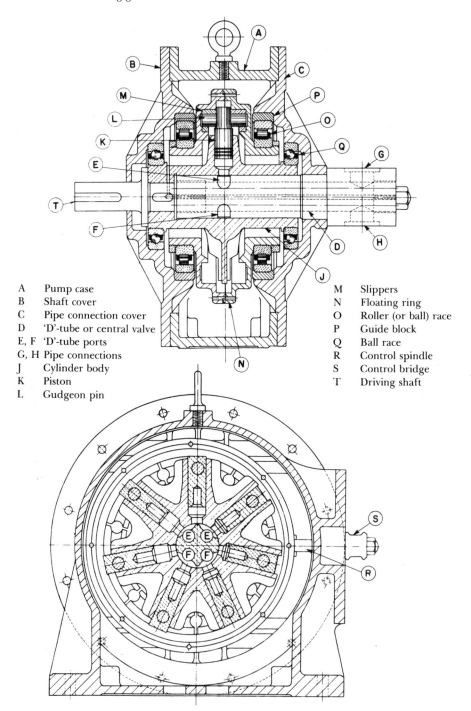

A	Pump case	M	Slippers
B	Shaft cover	N	Floating ring
C	Pipe connection cover	O	Roller (or ball) race
D	'D'-tube or central valve	P	Guide block
E, F	'D'-tube ports	Q	Ball race
G, H	Pipe connections	R	Control spindle
J	Cylinder body	S	Control bridge
K	Piston	T	Driving shaft
L	Gudgeon pin		

Figure 12.1 Hele-Shaw pump

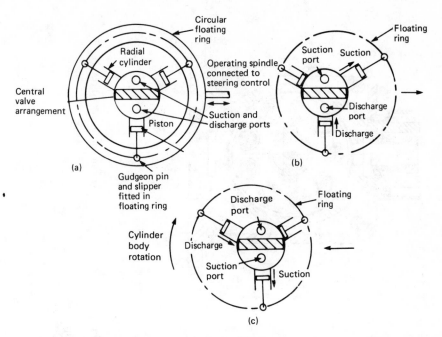

Figure 12.2 Hele-Shaw pump—operating principle

The swash plate and slipper pad designs are both axial cylinder pumps. The slipper pad is an improvement on the swash plate which provides higher pressure. An arrangement of a swash plate pump is shown in Figure 12.3. The driving shaft rotates the cylinder barrel, swash plate and pistons. An external trunnion (short shaft) enables the swash plate to be moved about its axis. The cylinders in the barrel are connected to ports which extend in an arc around the fixed port plate.

When the swash plate is vertical no pumping action takes place. When the swash plate is tilted pumping occurs, the length of stroke depending upon the angle of tilt. Depending upon the direction of tilt the ports will be either suction or discharge. This pump arrangement therefore offers the same flexibility in operation as the radial piston type.

Telemotor control

Telemotor control is a hydraulic control system employing a transmitter, a receiver, pipes and a charging unit. The transmitter, which is built into the steering wheel console, is located on the bridge and the receiver is mounted on the steering gear. The charging unit is located near to the receiver and the system is charged with a non-freezing fluid.

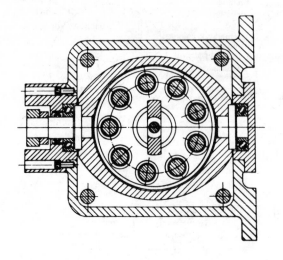

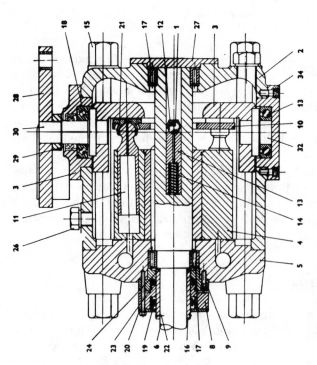

Figure 12.3 Swash plate pump

1 Steel ball	11 Piston
2 Pump body	12 Bridge piece
3 Tilt box (swash plate)	13 Plunger
4 Cylinder block	14 Spring
5 Valve plate	15 Case nuts and bolts
6 Mainshaft	16 Shaft sleeve
8 Oil seal housing	17 Needle bearing
9 Gland housing	18 Roller journal
10 Retracting plate	19 Oil seal

20 Retaining plate	32, 34 Bottom trunnion and cover
21 Slipper	
22 Circlip	
23, 24 O-rings	
26 Vent plug	
27 Roller bearing cap	
28 Control lever	
29 Oil seal	
30 Top trunnion and cover	

The telemotor system is shown in Figure 12.4. Two rams are present in the transmitter which move in opposite directions as the steering wheel is turned. The fluid is therefore pumped down one pipe line and drawn in from the other. The pumped fluid passes through piping to the receiver and forces the telemotor cylinder unit to move. The suction of fluid from the opposite cylinder enables this movement to take place. The cylinder unit has a control spindle connected to it by a pin. This control spindle operates the slipper ring or swash plate of the variable delivery pump. If the changeover pin is removed from the cylinder unit and inserted in the local handwheel drive then manual control of the

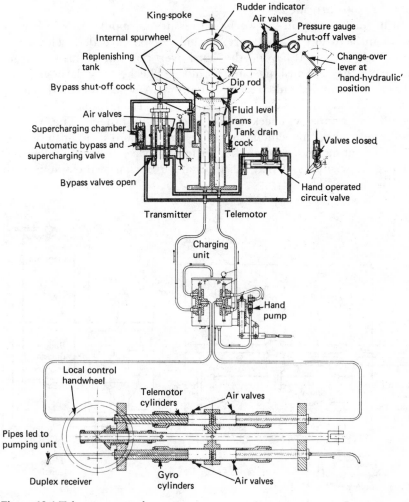

Figure 12.4 Telemotor control system

steering gear is possible. Stops are fitted on the receiver to limit movement to the maximum rudder angle required. The charging unit consists of a tank, a pump, and shut-off cocks for each and is fitted in the main piping between the transmitter and receiver.

In the transmitter a replenishing tank surrounds the rams, ensuring that air cannot enter the system. A bypass between the two cylinders opens as the wheel passes midships. Also at mid position the supercharging unit provides a pressure in the system which ensures rapid response of the system to a movement of the wheel. This supercharging unit also draws in replenishing fluid if required in the system, and provides a relief valve arrangement if the pressure is too high. Pressure gauges are connected to each main pipeline and air vent cocks are also provided.

In normal operation the working pressure of about 20 to 30 bar, or the manufacturer's given figure, should not be exceeded. The wheel should not be forced beyond the 'hard over' position as this will strain the gear. The replenishing tank should be checked regularly and any lubrication points should receive attention. Any leaking or damaged equipment must be repaired or replaced as soon as possible. The system should be regularly checked for pressure tightness. The rudder response to wheel movement should be checked and if sluggish or slow then air venting undertaken. If, after long service, air venting does not remove sluggishness, it may be necessary to recharge the system with new fluid.

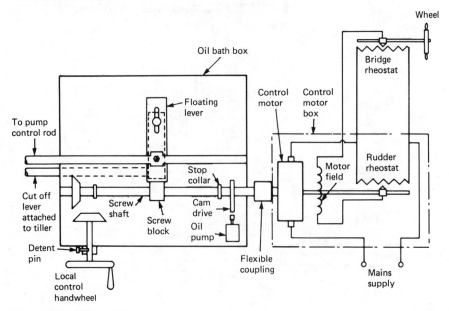

Figure 12.5(a) Control box

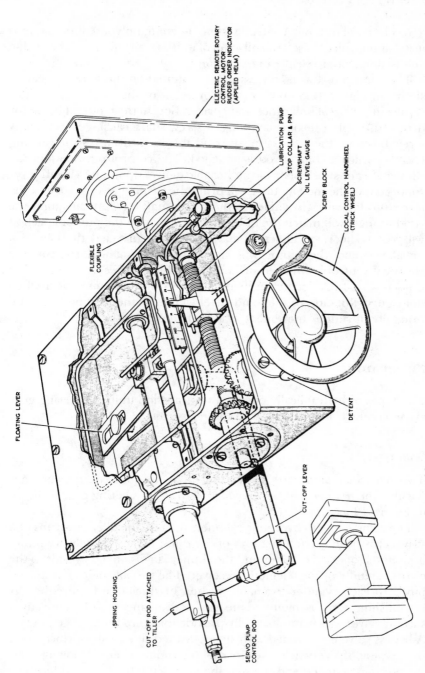

ELECTRIC REMOTE ROTARY
CONTROL MOTOR
RUDDER ORDER INDICATOR
(APPLIED HELM)

LUBRICATION PUMP
STOP COLLAR & PIN
SCREWSHAFT
OIL LEVEL GAUGE

SCREW BLOCK

LOCAL CONTROL HANDWHEEL
(TRICK WHEEL)

FLEXIBLE
COUPLING

FLOATING LEVER

DETENT

CUT-OFF LEVER

SPRING HOUSING

CUT-OFF ROD ATTACHED
TO TILLER

SERVO PUMP
CONTROL ROD

Figure 12.5(b) Arrangement of control box

Electrical control

The electrical remote control system is commonly used in modern installations since it uses a small control unit as transmitter on the bridge and is simple and reliable in operation.

The control box assembly, which is mounted on the steering gear, is shown in Figure 12.5 (a) and (b). Movement of the bridge transmitter results in electrical imbalance and current flow to the motor. The motor drives, through a flexible coupling, a screw shaft, causing it to turn. A screw block on the shaft is moved and this in turn moves the floating lever to which a control rod is attached. The control rod operates the slipper ring or swash plate of the variable delivery pump. A cut-off lever connected to the moving tiller will bring the floating lever pivot and the lever into line at right angles to the screw shaft axis. At this point the rudder angle will match the bridge lever angle and the pumping action will stop. The rotating screw shaft will have corrected the electrical imbalance and the motor will stop. For local manual control, the electrical control is switched off and a small handwheel is connected to the screw shaft. A detent pin holds the handwheel assembly clear when not in use. Rotation of the handwheel will move the floating lever and bring about rudder movement as already described.

Power units

Two types of hydraulically powered transmission units or steering gear are in common use, the *ram* and the *rotary vane*.

Ram type

Two particular variations, depending upon torque requirements, are possible the *two-ram* and the *four-ram*. A two-ram steering gear is shown in Figure 12.6.

The rams acting in hydraulic cylinders operate the tiller by means of a swivel crosshead carried in a fork of the rams. A variable delivery pump is mounted on each cylinder and the slipper ring is linked by rods to the control spindle of the telemotor receiver. The variable delivery pump is piped to each cylinder to enable suction or discharge from either. A replenishing tank is mounted nearby and arranged with non-return suction valves which automatically provide make-up fluid to the pumps. A bypass valve is combined with spring-loaded shock valves which open in the event of a very heavy sea forcing the rudder over. In moving over, the pump is actuated and the steering gear will return the rudder to its original position once the heavy sea has passed. A spring-loaded return

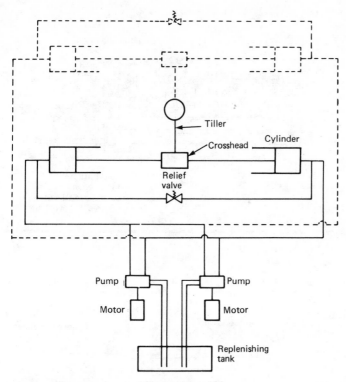

Figure 12.6(a) Diagrammatic arrangement of two-ram steering gear (additional items for four-ram system shown dotted)

linkage on the tiller will prevent damage to the control gear during a shock movement.

During normal operation one pump will be running. If a faster response is required, for instance in confined waters, both pumps may be in use. The pumps will be in the no-delivery state until a rudder movement is required by a signal from the bridge telemotor transmitter. The telemotor receiver cylinder will then move: this will result in a movement of the floating lever which will move the floating ring or slipper pad of the pump, causing a pumping action. Fluid will be drawn from one cylinder and pumped to the other, thus turning the tiller and the rudder. A return linkage or hunting gear mounted on the tiller will reposition the floating lever so that no pumping occurs when the required rudder angle is reached.

A four-ram steering gear is shown in Figure 12.7. The basic principles of operation are similar to the two-ram gear except that the pump will draw from two diagonally opposite cylinders and discharge to the other two. The four-ram arrangement provides greater torque and the

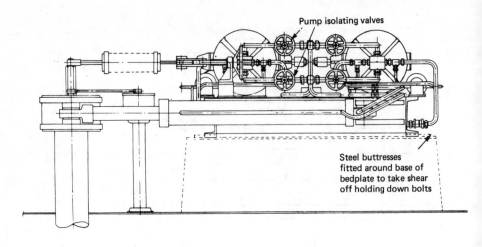

Pump isolating valves

Steel buttresses
fitted around base of
bedplate to take shear
off holding down bolts

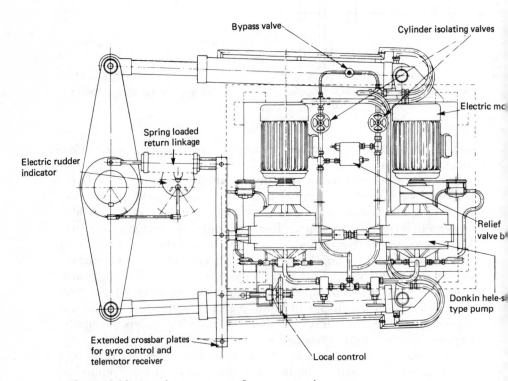

Bypass valve

Cylinder isolating valves

Electric mo

Spring loaded
return linkage

Electric rudder
indicator

Relief
valve b

Donkin hele-s
type pump

Extended crossbar plates
for gyro control and
telemotor receiver

Local control

Figure 12.6(b) Actual arrangement of two-ram steering gear

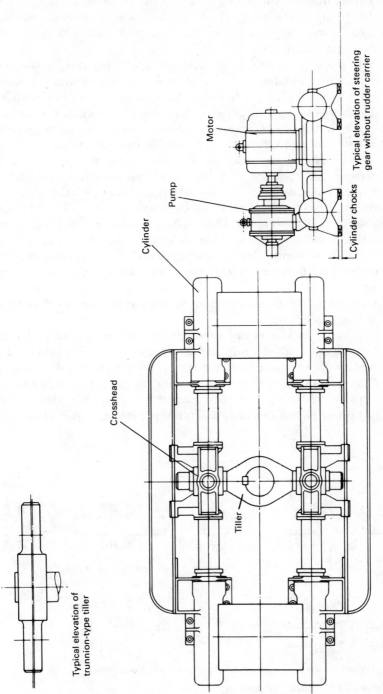

Motor

Pump

Cylinder

Crosshead

Tiller

Cylinder chocks

Typical elevation of steering
gear without rudder carrier

Typical elevation of
trunnion-type tiller

Figure 12.7 Four-ram steering gear—actual arrangement

flexibility of different arrangements in the event of component failure. Either pump can be used with all cylinders or with either the two port or two starboard cylinders. Various valves must be open or closed to provide these arrangements.

The use of a control valve block incorporating rudder shock relief valves, pump isolating valves, ram isolating and bypass valves, offers greater flexibility with a four-ram steering gear. In normal operation one pump can operate all cylinders. In an emergency situation the motor or a pair of hand pumps could be used to operate two port rams, two starboard rams, two forward rams or two after rams.

The crosshead arrangement on the four-ram type steering gear described incorporates what is known as the 'Rapson Slide'. This provides a mechanical advantage which increases with the angle turned through. The crosshead arrangement may use either a forked tiller or a round arm tiller (Figure 12.8). The round arm tiller has a centre crosshead which is free to slide along the tiller. Each pair of rams is joined so as to form a double bearing in which the trunnion arms of the crosshead are mounted. The straight line movement of the rams is thus converted into an angular tiller movement. In the forked tiller arrangement the ram movement is transferred to the tiller through swivel blocks.

To charge the system with fluid it is first necessary to fill each cylinder then replace the filling plugs and close the air cocks. The cylinder bypass valves should be opened and the replenishing tanks filled. The air vents on the pumps should be opened until oil discharges free of air, the pumps set to pump and then turned by hand, releasing air at the appropriate pair of cylinders and pumping into each pair of cylinders in

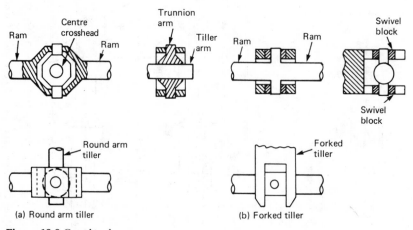

Figure 12.8 Crosshead arrangements

turn using the hand control mechanism. The motor should then be started up and, using the local hand control, operation of the steering gear checked. Air should again be released from the pressurised cylinders and the pumps through the appropriate vents.

During normal operation the steering gear should be made to move at least once every two hours to ensure self lubrication of the moving parts. No valves in the system, except bypass and air vent, should be closed. The replenishing tank level should be regularly checked and, if low, refilled and the source of leakage found. When not in use, that is, in port, the steering motors should be switched off. Also the couplings of the motors should be turned by hand to check that the pump is moving freely. If there is any stiffness the pump should be overhauled. As with any hydraulic system cleanliness is essential when overhauling equipment and only linen cleaning cloths should be used.

Rotary vane type

With this type of steering gear a vaned rotor is securely fastened onto the rudder stock (Figure 12.9). The rotor is able to move in a housing which is solidly attached to the ship's structure. Chambers are formed between the vanes on the rotor and the vanes in the housing. These chambers will vary in size as the rotor moves and can be pressurised since sealing strips are fitted on the moving faces. The chambers either side of the moving vane are connected to separate pipe systems or manifolds. Thus by supplying hydraulic fluid to all the chambers to the left of the moving vane and drawing fluid from all the chambers on the right, the rudder stock can be made to turn anti-clockwise. Clockwise movement will occur if pressure and suction supplies are reversed. Three vanes are usual and permit an angular movement of 70°: the vanes also act as stops limiting rudder movement. The hydraulic fluid is supplied by a variable delivery pump and control will be electrical, as described earlier. A relief valve is fitted in the system to prevent overpressure and allow for shock loading of the rudder.

All-electric steering

Steering gears which comprise electric control, electric power unit and electrical transmission, are of two types, the Ward–Leonard system and the Direct Single Motor system. Both types have a geared-down motor drive via a pinion to a toothed quadrant.

A Ward–Leonard arrangement is shown diagrammatically in Figure 12.10. A continuously running motor-generator set has a directly coupled exciter to provide the field current of the generator. The

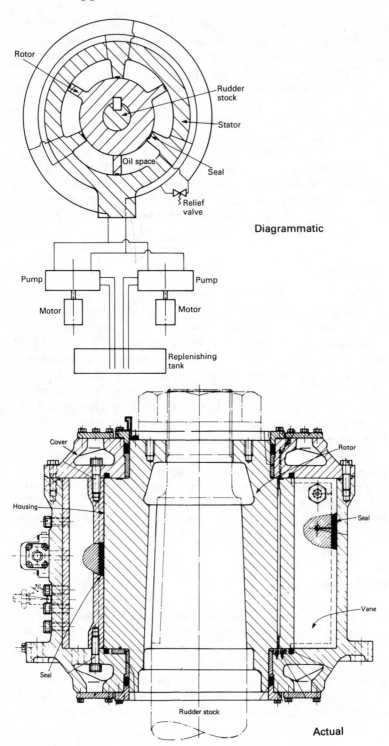

Figure 12.9(a) Rotary vane steering gear

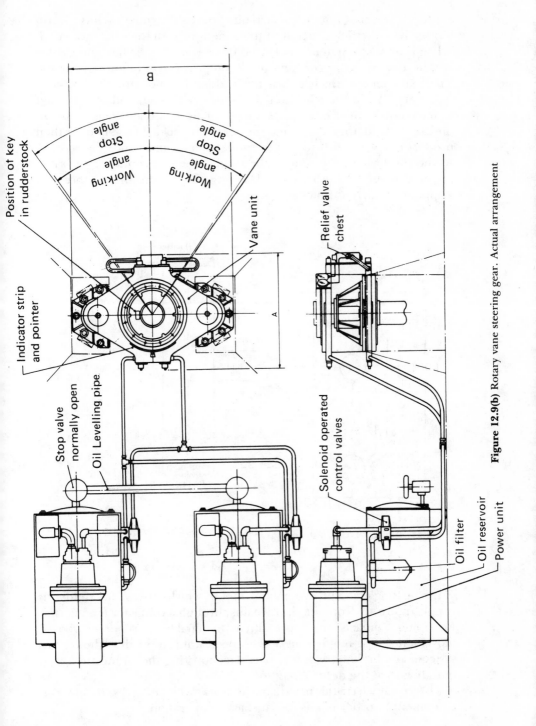

Figure 12.9(b) Rotary vane steering gear. Actual arrangement

Position of key in rudderstock

Indicator strip and pointer

Stop valve normally open

Oil Levelling pipe

Working angle

Stop angle

Stop angle

Working angle

Vane unit

Relief valve chest

Solenoid operated control valves

Oil filter

Oil reservoir

Power unit

A

B

exciter field is part of a control circuit, although in some circuits control is directly to the field current of the generator with the exciter omitted. When the control system is balanced there is no exciter field, no exciter output and no generator output, although it is continuously running. The main motor which drives the rudder has no input and thus is stationary. When the wheel on the bridge is turned, and the rheostat contact moved, the control system is unbalanced and a voltage occurs in the exciter field, the exciter, and the generator field. The generator then produces power which turns the rudder motor and hence the rudder. As the rudder moves it returns the rudder rheostat contact to the same position as the bridge rheostat, bringing the system into balance and stopping all current flow.

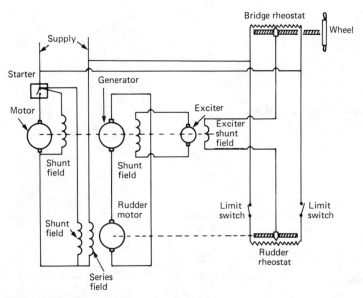

Figure 12.10 Ward–Leonard steering gear

In the single motor system the motor which drives the rudder is supplied directly from the ship's mains through a contactor type starter. Reversing contacts are also fitted to enable port or starboard movements. The motor runs at full speed until stopped by the control system, so a braking system is necessary to bring the rudder to a stop quickly and at the desired position.

The usual electrical maintenance work will be necessary on this equipment in order to ensure satisfactory operation.

Twin system steering gears

To meet the automatic changeover within the 45 seconds required for tankers of 10 000 ton gross tonnage and above, a number of designs are available. Two will be described, one for a ram type steering gear and one for a rotary vane type steering gear. In each case two independent systems provide the power source to move the tiller, the failure of one resulting in a changeover to the other. The changeover is automatic and is achieved within 45 seconds.

The ram type steering gear arrangement is shown diagrammatically in Figure 12.11. A simple automatic device monitors the quantity of oil in the circuit. Where a failure occurs in one of the systems it is located and

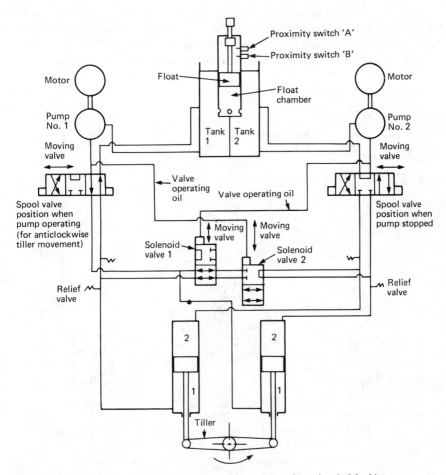

Figure 12.11 Ram type twin circiut system—pump 1 running, circuit 2 leaking

that circuit is isolated. The other system provides uninterrupted steering and alarms are sounded and displayed.

Consider pump 1 in operation and pump 2 placed on automatic reserve by the selector switch. If a leak develops in circuit 2 the float chamber oil level will fall and proximity switch A on the monitor will be activated to close the solenoid valve 2 which isolates circuit 2 and bypasses the cylinders in that circuit. An alarm will also be given. If the leak is in circuit 1 however, the float chamber oil level will fall further until proximity switch B is activated. This will cut off the power supply to motor 1 and solenoid valve 1 and connect the supply to motor 2 and solenoid valve 2, thus isolating circuit 1. If pump 2 were running and pump 1 in reserve, a similar changeover would occur. While a two cylinder system has been described this system will operate equally well with four double acting cylinders.

An arrangement based on a rotary vane type steering gear is shown in Figure 12.12. This system involves the use of only one actuator but it is directly fitted to a single tiller and rudderstock and therefore complete duplication of the system does not occur anyway. Self closing lock valves are provided in the two independent hydraulic circuits which operate

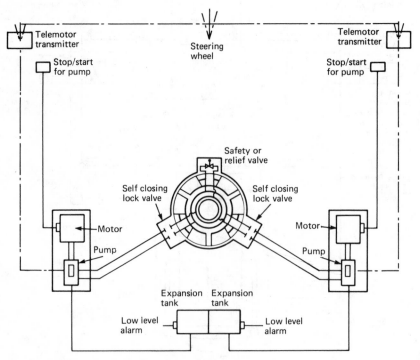

Figure 12.12 Rotary-vane twin circuit system

the actuator. The self closing valves are fitted on the inlet and outlet ports of the actuator and open under oil pressure against the action of a spring. Where an oil pressure loss occurs in one circuit the valves will immediately close under the action of their springs. A low tank level alarm will sound and the other pump can be started. This pump will build up pressure, open the valves on its circuit and the steering gear can immediately operate.

Steering gear testing

Prior to a ship's departure from any port the steering gear should be tested to ensure satisfactory operation. These tests should include:

1. Operation of the main steering gear.
2. Operation of the auxiliary steering gear or use of the second pump which acts as the auxiliary.
3. Operation of the remote control (telemotor) system or systems from the main bridge steering positions.
4. Operation of the steering gear using the emergency power supply.
5. The rudder angle indicator reading with respect to the actual rudder angle should be checked.
6. The alarms fitted to the remote control system and the steering gear power units should be checked for correct operation.

During these tests the rudder should be moved through its full travel in both directions and the various equipment items, linkages, etc., visually inspected for damage or wear. The communication system between the bridge and the steering gear compartment should also be operated.

_____Chapter 13_____

Fire fighting and safety

Fire is a constant hazard at sea. It results in more total losses of ships than any other form of casualty. Almost all fires are the result of negligence or carelessness.

Combustion occurs when the gases or vapours given off by a substance are ignited: it is the gas given off that burns, not the substance. The temperature of the substance at which it gives off enough gas to continue burning is known as the 'flash point'.

Fire is the result of a combination of three factors:

1. A substance that will burn.
2. An ignition source.
3. A supply of oxygen, usually from the air.

These three factors are often considered as the sides of the *fire triangle*. Removing any one or more of these sides will break the triangle and result in the fire being put out. The complete absence of one of the three will ensure that a fire never starts.

Fires are classified according to the types of material which are acting as fuel. These classifications are also used for extinguishers and it is essential to use the correct classification of extinguisher for a fire, to avoid spreading the fire or creating additional hazards. The classifications use the letters A, B, C, D and E.

Class A Fires burning wood, glass fibre, upholstery and furnishings.
Class B Fires burning liquids such as lubricating oil and fuels.
Class C Fires burning gas fuels such as liquefied petroleum gas.
Class D Fires burning combustible metals such as magnesium and aluminium.
Class E Fires burning any of the above materials together with high voltage electricity.

Many fire extinguishers will have multiple classifications such as A, B and C.

Fire fighting at sea may be considered in three distinct stages, *detection*—locating the fire; *alarm*—informing the rest of the ship; and *control*—bringing to bear the means of extinguishing the fire.

Detection

The use of fire detectors is increasing, particularly with the tendency to reduced manning and unmanned machinery spaces. A fire, if detected quickly, can be fought and brought under control with a minimum of damage. The main function of a fire detector is therefore to detect a fire as quickly as possible; it must also be reliable and require a minimum of attention. An important requirement is that it is not set off by any of the normal occurrences in the protected space, that is it must be appropriately sensitive to its surroundings. Three phenomena associated with fire are used to provide alarms: these are *smoke, flames* and *heat*.

The smoke detector makes use of two ionisation chambers, one open to the atmosphere and one closed (Figure 13.1). The fine particles or aerosols given off by a fire alter the resistance in the open ionisation chamber, resulting in the operation of a cold cathode gas-filled valve. The alarm sounds on the operation of the valve to give warning of a fire. Smoke detectors are used in machinery spaces, accommodation areas and cargo holds.

Flames, as opposed to smoke, are often the main result of gas and liquid fires and flame detectors are used to protect against such hazards. Flames give off ultra-violet and infra-red radiation and detectors are available to respond to either. An infra-red flame detector is shown in Figure 13.2. Flame detectors are used near to fuel handling equipment in the machinery spaces and also at boiler fronts.

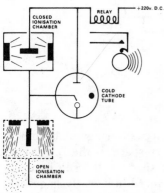

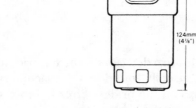

Figure 13.1 Smoke detector

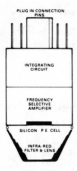

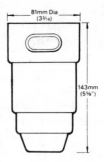

Figure 13.2 Infra-red flame detector

Heat detectors can use any of a number of principles of operation, such as liquid expansion, low melting point material or bimetallic strips. The most usual detector nowadays operates on either a set temperature rise or a rate of temperature rise being exceeded. Thus an increase in temperature occurring quickly could set off the alarm before the set temperature was reached. The relative movement of two coiled bimetallic thermostats, one exposed and one shielded, acts as the detecting element (Figure 13.3). Heat detectors are used in places such as the galley and laundry where other types of detector would give off false alarms.

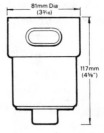

Figure 13.3 Heat detector

Alarm

Associated with fire detectors is the electric circuit to ring an alarm bell. This bell will usually sound in the machinery space, if the fire occurs there, and also on the bridge. Fires in other spaces will result in alarm bells sounding on the bridge. Any fire discovered in its early stages will require the finder to give the alarm and or make the decision to deal with it himself if he can. Giving the alarm can take many forms such as shouting 'Fire', banging on bulkheads or any action necessary to attract attention. It is necessary to give an alarm in order to concentrate resources and effort quickly onto the fire, even if the fire must be left to burn for a short time unchecked.

Control

Two basically different types of equipment are available on board ship for the control of fires. These are small portable extinguishers and large fixed installations. The small portable extinguishers are for small fires which, by prompt on-the-spot action, can be rapidly extinguished. The fixed installation is used when the fire cannot be fought or restrained by portable equipment or there is perhaps a greater danger if associated areas were to be set on fire. The use of fixed installations may require evacuation of the area containing the fire which, if it is the machinery space, means the loss of effective control of the ship. Various types of both portable and fixed fire fighting equipment are available.

Fire fighting equipment

Portable extinguishers

There are four principal types of portable extinguisher usually found on board ship. These are the *soda-acid, foam, dry powder* and *carbon dioxide* extinguishers.

Soda-acid extinguisher

The container of this extinguisher holds a sodium bicarbonate solution. The screw-on cap contains a plunger mechanism covered by a safety guard. Below the plunger is a glass phial containing sulphuric acid (Figure 13.4). When the plunger is struck the glass phial is broken and the acid and sodium bicarbonate mix. The resulting chemical reaction produces carbon dioxide gas which pressurises the space above the liquid forcing it out through the internal pipe to the nozzle. This extinguisher is used for Class A fires and will be found in accommodation areas.

Foam extinguisher—chemical √

The main container is filled with sodium bicarbonate solution and a long inner polythene container is filled with aluminium sulphate (Figure 13.5(a)). The inner container is sealed by a cap held in place by a plunger. When the plunger is unlocked by turning it, the cap is released.

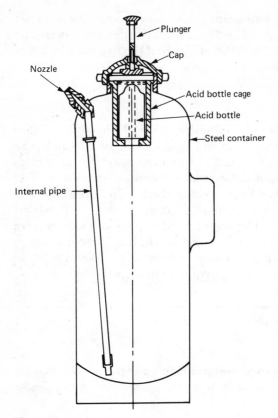

Figure 13.4 Soda-acid extinguisher

The extinguisher is then inverted for the two liquids to mix. Carbon dioxide is produced by the reaction which pressurises the container and forces out the foam.

Foam extinguisher—mechanical

The outer container in this case is filled with water. The central container holds a carbon dioxide charge and a foam solution (Figure 13.5(b)). A plunger mechanism with a safety guard is located above the central container. When the plunger is depressed the carbon dioxide is released and the foam solution and water mix. They are then forced out through a special nozzle which creates the mechanical foam. This extinguisher has an internal pipe and is operated upright.

Foam extinguishers are used on Class B fires and will be located in the vicinity of flammable liquids.

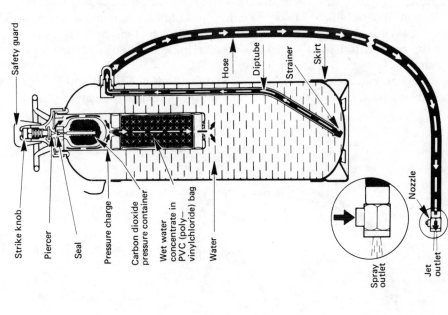

Figure 13.5(b) Foam fire extinguishers—mechanical foam

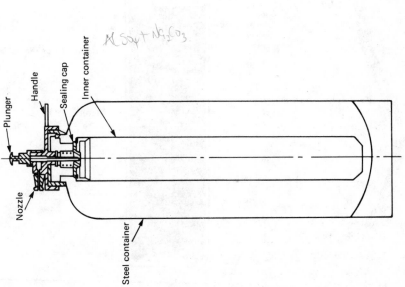

Figure 13.5(a) Foam fire extinguishers—chemical foam

Carbon dioxide extinguisher

A very strong container is used to store liquid carbon dioxide under pressure (Figure 13.6). A central tube provides the outlet passage for the carbon dioxide which is released either by a plunger bursting a disc or a valve operated by a trigger. The liquid changes to a gas as it leaves the extinguisher and passes through a swivel pipe or hose to a discharge horn.

Carbon dioxide extinguishers are mainly used on Class B and C fires and will be found in the machinery space, particularly near electrical

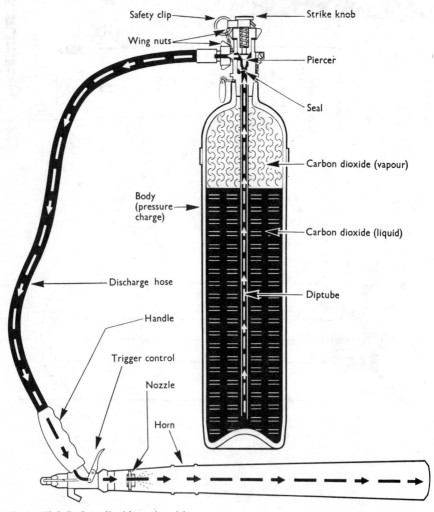

Figure 13.6 Carbon dioxide extinguisher

equipment. The carbon dioxide extinguisher is not permitted in the accommodation since, in a confined space, it could be lethal.

Dry powder extinguishers

The outer container contains sodium bicarbonate powder. A capsule of carbon dioxide gas is located beneath a plunger mechanism in the central cap (Figure 13.7). On depressing the plunger the carbon dioxide gas forces the powder up a discharge tube and out of the discharge nozzle.

The dry powder extinguisher can be used on all classes of fire but it has no cooling effect. It is usually located near electrical equipment in the machinery space and elsewhere on the ship.

Maintenance and testing

All portable extinguishers are pressure vessels and must therefore be regularly checked.

The soda-acid and foam extinguisher containers are initially tested to 25 bar for five minutes and thereafter at four-yearly intervals to 20 bar.

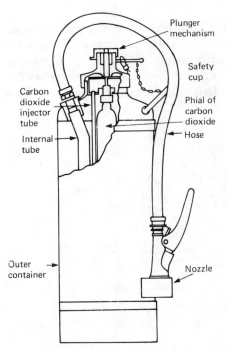

Figure 13.7 Dry powder extinguisher

The carbon dioxide extinguisher is tested to 207 bar initially every 10 years and after two such tests, every five years. The dry powder extinguisher is tested to 35 bar once every four years.

Most extinguishers should be tested by discharge over a period of one to five years, depending on the extinguisher type, e.g. soda-acid and dry powder types 20% discharged per year, foam types 50% discharged per year. Carbon dioxide extinguishers should be weighed every six months to check for leakage.

Where practicable the operating mechanisms of portable extinguishers should be examined every three months.

Any plunger should be checked for free movement, vent holes should be clear and cap threads lightly greased. Most extinguishers with screw-on caps have a number of holes in the threaded region. These are provided to release pressure before the cap is taken off: they should be checked to be clear.

Fixed installations

A variety of different fixed fire fighting installations exist, some of which are specifically designed for certain types of ship. A selection of the more general installations will now be outlined.

Fire main

A sea water supply system to fire hydrants is fitted to every ship (Figure 13.8). Several pumps in the engine room will be arranged to supply the system, their number and capacity being dictated by legislation (Department of Transport for UK registered vessels). An emergency fire pump will also be located remote from the machinery space and with independent means of power.

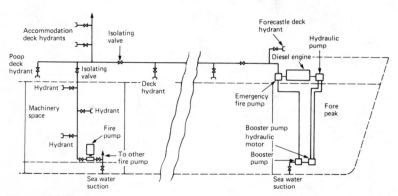

Figure 13.8 Fire main

A system of hydrant outlets, each with an isolating valve, is located around the ship, and hoses with appropriate snap-in connectors are strategically located together with nozzles. These nozzles are usually of the jet/spray type providing either type of discharge as required. All the working areas of the ship are thus covered, and a constant supply of sea water can be brought to bear at any point to fight a fire.

While sea water is best used as a cooling agent in fighting Class A fires it is possible, if all else fails, to use it to fight Class B fires. The jet/spray nozzle would be adjusted to provide a fine water spray which could be played over the fire to cool it without spreading.

An international shore connection is always carried on board ship. This is a standard size flange which is fitted with a coupling suitable for the ship's hoses. The flange is slotted in order to fit any shore-side fire main and enable water to be brought on board a ship lying alongside.

Automatic water spray

The automatic spray or sprinker system provides a network of sprinkler heads throughout the protected spaces. This system may be used in accommodation areas, and in machinery spaces with certain variations in the equipment used and the method of operation.

The accommodation areas are fitted with sprinkler heads which both detect and extinguish fires. The sprinkler head is closed by a quartzoid bulb which contains a liquid that expands considerably on heating (Figure 13.9). When excessively heated the liquid expands, shatters the bulb and water will issue from the sprinkler head. A deflector plate on the sprinkler head causes the water to spray out over a large area.

The water is supplied initially from a tank pressurised by compressed air (Figure 13.10(a)). Once the tank pressure falls, as a sprinkler issues water, a salt water pump cuts in automatically to maintain the water supply as long as is necessary. The system is initially charged with fresh water to reduce corrosion effects.

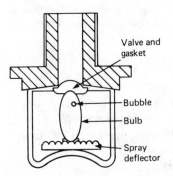

Valve and gasket

Bubble

Bulb

Spray deflector

Figure 13.9 Sprinkler head

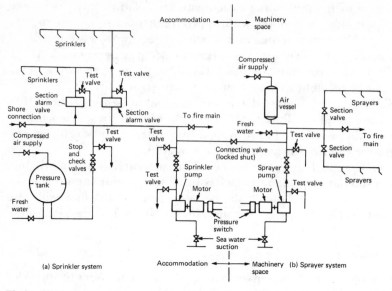

Figure 13.10 Automatic water spray systems, (a) sprinkler system; (b) sprayer system

The complete installation is divided into several sections, each containing about 150 to 200 sprinklers and having an alarm valve. When one or more sprinklers operate water flows through the section valve and sounds an alarm and also provides a visual display identifying the section containing the fire.

In the machinery space the sprinkler heads are known as 'sprayers' and have no quartzoid bulb. Also the section valves are manually operated to supply water to the sprayers (Figure 13.10(b)). The system is pressurised by compressed air with a salt water pump arranged to cut in automatically if the pressure drops. The accommodation and machinery space systems may be combined by a valve which is normally kept locked shut.

The system should be regularly checked by creating fault conditions at the various section control valves by opening a test valve, and checking for audible and visual alarms.

Foam systems

Foam spreading systems are designed to suit the particular ship's requirements with regard to quantity of foam, areas to be protected, etc. Mechanical foam is the usual substance used, being produced by mixing foam making liquid with large quantities of water. Violent agitation of the mixture in air creates air bubbles in the foam.

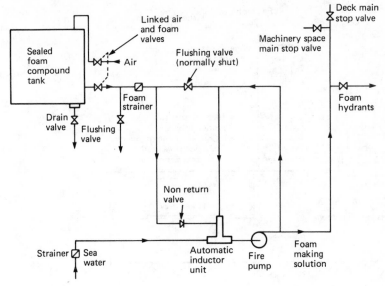

Figure 13.11 Foam induction system

An automatic foam induction system is shown in Figure 13.11. The automatic inductor unit ensures the correct mixing of water and foam compound which is then pumped as the foam making solution to the hydrants for use. The foam compound tank is sealed to protect the contents from deterioration and has linked compound supply and air vent valves. To operate the system these two linked valves are opened and the fire pump started. Foam mixing is carefully metered by the automatic inductor unit. The fire pump and compound tank must be located outside the protected space.

High-expansion foam systems are also available where a foam generator produces, from foam concentrate and sea water, a thousand times the quantity of foam. The generator blows air through a net sprayed with foam concentrate and water. The vastly expanded foam is then ducted away to the space to be protected. The foam is an insulator and an absorber of radiant heat; it also excludes oxygen from the fire.

Carbon dioxide flooding

A carbon dioxide flooding system is used to displace the oxygen in the protected space and thus extinguish the fire. The carbon dioxide is stored as a liquid under pressure in cylinders.

The volume of space to be protected determines the number of cylinders required. A common battery of cylinders may be used to protect both cargo holds and machinery space.

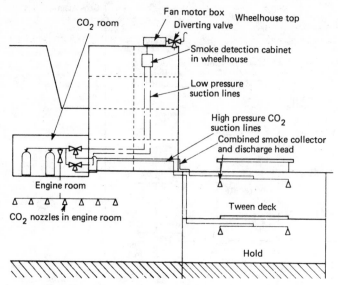

Figure 13.12 Carbon dioxide detection and flooding system

The cargo space system is normally arranged for smoke detection, alarm and carbon dioxide flooding (Figure 13.12).

Small air sampling pipes from the individual cargo holds are led into a cabinet on the bridge. Air is drawn from each hold by a small fan and each pipe is identified for its particular hold. If smoke is drawn into the cabinet from one of the holds it will set off an alarm. The smoke is also passed into the wheelhouse where it can be detected by personnel on watch.

The location of the fire can be identified in the cabinet and the hold distribution valve below the cabinet is operated. This valve shuts off the sampling pipe from the cabinet and opens it to the carbon dioxide main leading from the cylinder battery. A chart will indicate the number of cylinders of gas to be released into the space and this is done by a hand operated lever.

The machinery space system is designed to quickly discharge the complete battery of cylinders. Before the gas is released the space must be clear of personnel and sealed against entry or exhausting of air.

The discharge valve is located in a locked cabinet, with the key in a glass case nearby. Opening the cabinet sounds an alarm to warn personnel of the imminent discharge of the gas. The discharge valve is opened and an operating lever pulled.

The operating lever opens two gas bottles which pressurise a gang release cylinder that, in turn, moves an operating cable to open all the bottles in the battery. The carbon dioxide gas then quickly floods the machinery space, filling it to 30% of its volume in two minutes or less.

The air sampling system can be checked when the holds are empty by using a smoking rag beneath a sampling point. Flow indicators, usually small propellers, are fitted at the outlet points of the smoke detecting pipes as a visual check and an assurance that the pipes are clear. To check for leakage the gas cylinders can be weighed or have their liquid levels measured by a special unit.

Inert gas

Inert gases are those which do not support combustion and are largely nitrogen and carbon dioxide. Large quantities suitable for fire extinguishing can be obtained by burning fuel in carefully measured amounts or by cleaning the exhaust gases from a boiler.

Inert gas generator

The inert gas generator (Figure 13.13) burns fuel in designed quantities to produce perfect combustion. This provides an exhaust gas which is largely nitrogen and carbon dioxide with a very small oxygen content. The exhaust gases pass to a cooling and washing chamber to remove sulphur and excess carbon. The washed or scrubbed exhaust gas is now inert and passes to a distribution system for fire extinguishing. The complete unit is arranged to be independently operated in order to supply inert gas for as long as the fuel supply lasts.

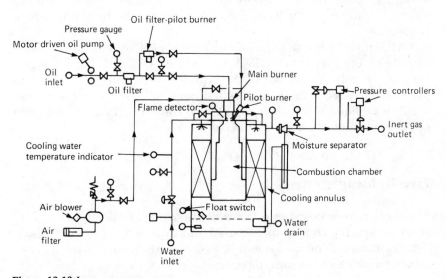

Figure 13.13 Inert gas generator

Funnel gas inerting

A system much used on tankers where boiler exhaust gases are cleaned and inerted is shown in Figure 13.14. The exhaust gas is cleaned in a scrubbing tower, dried and filtered before being passed to the deck mains for distribution. The gas will contain less than 5% oxygen and is therefore considered inert. It is distributed along the deck pipes by fans and passes into the various cargo tanks. Seals in the system act as non-return valves to prevent a reverse flow of gas.

The inert gas is used to blanket the oil cargo during discharging operations. Empty tanks are filled with gas and the inert gas is blown out when oil is loaded.

Inert gas-producing units have the advantage of being able to continuously produce inert gas. A bottle storage system, such as carbon dioxide flooding, is a 'one-shot' fire extinguisher which leaves a ship unprotected until further gas supplies can be obtained.

Halon system

Halon 1301 (BTM) and Halon 1211 (BCF) are two halogenated hydrocarbon gases with special fire extinguishing properties. Unlike other extinguishing agents which cool the fire or displace oxygen the Halon gases inhibit the actual flame reaction. As a result of its low vapour pressure when liquefied Halon can be stored in low-pressure containers. Alternatively if a standard carbon dioxide cylinder is used then approximately three times as much gas can be stored. An additional advantage is that the atmosphere in a Halon flooded space is not toxic, although some highly irritant gases are produced in the extinguishing process.

A Halon storage system would be very similar to one using carbon dioxide except that fewer cylinders would be required. The liquefied Halon is usually pressurised in the cylinders with nitrogen in order to increase the speed of discharge. Bulk storage tanks of Halon gas are also used with cylinders of carbon dioxide and compressed air being used to operate the control system and expel the gas.

Fire fighting strategy

Fighting a fire on board ship may amount to a life or death struggle; to enter into such a conflict unprepared and unarmed is to invite failure. The 'armaments' or equipment available have been described. Now comes the matter of being prepared.

A basic strategy should be followed in all fire fighting situations. This

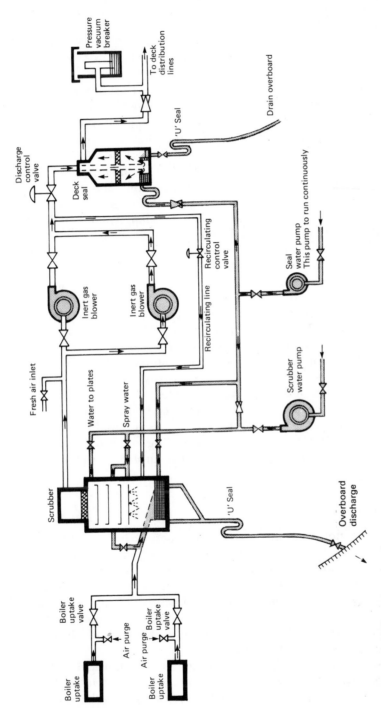

Figure 13.14 Funnel gas inerting system

will involve four distinct aspects, which are *locating, informing, containing* and finally *extinguishing* a fire.

A fire may be located by detection devices fitted in the various spaces in a ship or simply by smelling or seeing smoke. Alert personnel, whether on watch or not, should always be conscious of the danger of fire and the signs which indicate it. Certain areas are more liable to outbreaks of fire and these should be regularly visited or checked upon.

Once detected the presence of a fire must be made known quickly to as many people as possible. It is essential therefore that the bridge is informed of the location and extent of the fire. A small fire might reasonably be immediately tackled by the finder but attempts should be made whilst fighting the fire to attract attention. Shouting 'Fire', banging on bulkheads, deliberately setting off equipment alarms in the vicinity, all are possible means of attracting attention. Anyone finding a fire must decide whether to fight it immediately or whether to leave it and inform others first. The more people who know of a fire the greater the efforts that can be brought to bear upon it. If in doubt—inform!

Ships are built to contain fires in the space where they begin. Fire resisting bulkheads and decks are positioned at appropriate distances in order to limit the spread of fire, and it remains for fire fighting personnel to ensure that these barriers are secure whilst attempting to fight the fire. All doors and openings should be closed, all ventilation and exhaust fans stopped, and flammable material isolated from the space. It should be remembered that a fire exists in three dimensions and therefore has six sides, so it must be contained on six sides.

A small fire can usually be easily extinguished but it can also quickly become a big fire, so the fire extinguishing must be rapid if it is to be effective. Fire fighting strategy will vary according to the location of the fire. The various areas and their particular problems will now be examined.

Accommodation

The accommodation areas will be made up almost exclusively of Class A material requiring the use of water or soda-acid type extinguishers. Electrical circuits however should be isolated before directing quantities of water into an accommodation area. All ventilation and exhaust fans must be stopped and fire flaps closed. If hoses are employed a water spray should be used in order to achieve the maximum cooling effect. The accommodation will no doubt fill with smoke and therefore breathing apparatus should be available.

The galley area presents a somewhat different fire hazard. Here Class B materials, such as cooking oil, fat or grease, will be present requiring the use of foam, dry powder or carbon dioxide extinguishers. A fire blanket quickly spread over burning cooking utensils could extinguish a potentially dangerous fire.

Machinery spaces

Machinery space fires will involve mainly Class B material requiring the use of foam type extinguishers. Only the smallest of fires should be tackled with hand extinguishers. The alarm should be quickly given and the bridge informed. The ventilation fans should be stopped and fire flaps closed. Any oil tanks close to the fire should be closed off and kept cool by hosing with water. Foam-making equipment should be used on the fire and foam spread over the tank tops and bilges. Water spray can also be used to cool the surroundings of the fire, but a water jet should not be used in the machinery space since it will move any burning oil around and subsequently spread the fire. Only if the situation becomes hopeless should the space be evacuated and gas flooding used. The machinery space contains most of the fire fighting equipment as well as the propulsion machinery. If it is vacated then control of the situation is lost to a 'one-shot' attempt at gas flooding.

If evacuation is decided upon all personnel must be made aware of the decision. The space must then be completely sealed against the entry or exit of air and all oil supplies isolated at the tank valves. When all these matters have been attended to, the flooding gas can be admitted and, if the surrounding bulkheads hold to contain the fire, it will quickly go out. Cooling of the boundary bulkheads should continue from outside the space whilst flooding is taking place.

When the extinguished fire has been left long enough to cool down the space can be re-entered. This should be done from the tunnel, if there is one, or the lowest point remote from the seat of the fire. Engineers wearing breathing apparatus may now enter, taking water spray hoses with them to cool down any hot surfaces. Cooling and smoke dispersal are the first priorities to provide an atmosphere in which others can operate and gradually bring the machinery back into service. Where a machinery space fire involves electrical equipment then only dry powder or carbon dioxide extinguishers can be used until the equipment is isolated.

Cargo spaces

Where a fire occurs in a cargo hold with a smoke detection and carbon dioxide flooding system fitted, the procedure is straightforward and has already been described. It is essential to ensure before flooding that all air entry and exit points are closed by fire dampers and all fans are stopped.

Oil tankers with their cargo tanks full or empty present a potentially serious fire hazard. A fire occurring in a cargo tank will doubtless lead to an explosion or an explosion will lead to fire. The rapid use of foam making equipment, the cooling of surrounding areas and the isolation of the fire should immediately take place.

The prevention of fire and explosion conditions is the main prerequisite with oil tankers. With reference to hydrocarbon vapours, such as those present in oil tanks, the diagram shown in Figure 13.15 should be considered. The relative proportions of hydrocarbon vapour and oxygen necessary for a fire or explosion are shown. By keeping the tank atmosphere outside of the flammable limits, no fire or explosion can occur. It is usual practice to inert the tank atmosphere by displacing the oxygen with an inert gas and thus effectively prevent a fire or explosion. The inert gas producing systems have already been described.

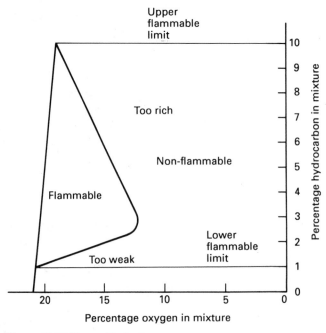

Figure 13.15 Flammable limits

Training and awareness

Where is the nearest fire extinguisher? What type is it? How is it operated? At any position in the ship these questions should be asked and answered. Knowing how to operate any extinguisher just by looking at it will indicate some degree of training and an awareness of the fire defences.

Fire drills are often referred to as 'Board of Trade Sports', but they merit a more sober attitude than they receive. Practices are useful and should be seriously undertaken. Equipment should be tried and tested to ensure that it works and is ready when needed. Regular maintenance should take place on extinguishers, fire pumps, hydrants, hoses, etc. All engineers should be familiar with recharging and overhauling extinguishers and those in charge should make sure it is regularly done. The statutory surveys do much to ensure that equipment is ready for use but the one year period between leaves a lot of time for neglect.

Breathing apparatus

Many fire fighting situations may require the use of some form of breathing apparatus. The use of such equipment will ensure a supply of oxygen to the wearer so that he can perform his particular tasks in safety. Two basic types are in use—the smoke helmet and the self-contained unit using air cylinders.

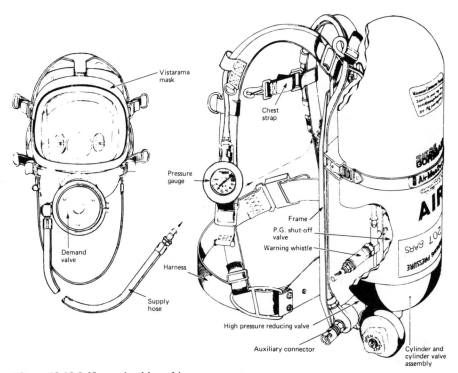

Figure 13.16 Self-contained breathing apparatus

The smoke helmet arrangement uses a helmet which covers the head and is connected to an air hose. A hand operated pump or bellows supplies the air. A system of signals between user and supplier must be arranged to ensure safe, correct operation.

The self-contained unit consists of one or two cylinders of compressed air kept in a harness which is carried on the back (Figure 13.16). The high pressure air is fed through a reducing valve and then to a demand valve. The demand valve is fitted into a face mask and supplies air to meet the breathing requirements of the wearer. A non-return valve permits breathing out to atmosphere. A warning whistle sounds when the air pressure falls to a low value. A standard cylinder will allow for about 20 to 30 minutes' operation.

Safe working practices

Accidents are usually the result of carelessness, mistakes, lack of thought or care, and often result in injury. Consideration will now be given to avoiding accidents, largely by the adoption of safe working practices.

Working clothes should be chosen with the job and its hazards in mind. They should fit fairly closely with no loose flaps, straps or ragged pockets. Clothing should cover as much of the body as possible and a stout pair of shoes should be worn. Neck chains, finger rings and wristwatches should not be worn, particularly in the vicinity of rotating machinery. Where particular hazards are present appropriate protection, such as goggles or ear muffs, should be worn.

When overhauling machinery or equipment it must be effectively isolated from all sources of power. This may involve unplugging from an electrical circuit, the removal of fuses or the securing open of circuit breakers. Suction and discharge valves of pumps should be securely closed and the pump casing relieved of pressure. Special care should be taken with steam-operated or steam-using equipment to ensure no pressure build-up can occur.

When lifting equipment during overhaul, screw-in eye bolts should be used where possible. These should be fully entered up to the collar and the threads on the eyebolt and in the equipment should be in good condition. Any lifting wires should be in good condition without broken strands or sharp edges.

Before any work is done on the main engine, the turning gear should be engaged and a warning posted at the control position. Lubricating oil in the working area should be cleaned up and where necessary suitable staging erected. The turning gear should be made inoperative if not required during the overhaul. Where it is used, care must be taken to ensure all personnel are clear before it is used.

Where overhead work is necessary suitable staging should be provided and adequately lashed down. Staging planks should be examined before use and where suspect discarded. Where ladders are used for access they must be secured at either end. Personnel working on staging should take care with tools and store them in a container.

Boiler blowbacks can cause serious injury and yet with care can usually be avoided. The furnace floor should be free of oil and burners regularly checked to ensure that they do not drip, particularly when not in use. The manufacturer's instructions should be followed with regard to lighting up procedures. Generally this will involve blowing through the furnace (purging) with air prior to lighting up. The fuel oil must be at the correct temperature and lit with a torch. If ignition does not immediately occur the oil should be turned off and purging repeated before a second attempt is made. The burner should be withdrawn and examined before it is lit.

Entry into an enclosed space should only take place under certain specified conditions. An enclosed space, such as a duct keel, a double bottom tank, a cofferdam, boiler, etc. cannot be assumed to contain oxygen. Anyone requiring to enter such a space should only do so with the permission of a responsible officer. The space should be well ventilated before entry takes place and breathing apparatus taken along: it should be used if any discomfort or drowsiness is felt. Another person should remain at the entrance to summon assistance if necessary, and there should be a means of communication arranged between the person within the space and the attendant. Lifelines and harness should be available at the entrance to the space. The attendant should first raise the alarm where the occupant appears in danger but should not enter the space unless equipped with breathing apparatus.

Training in the use of safety equipment and the conduct of rescues is essential for all personnel involved.

Chapter 14
Electrical equipment

The complete electrical plant on board ship is made up of power generation equipment, a distribution system and the many power utilising devices. Electricity is used for the motor drive of many auxiliaries and also for deck machinery, lighting, ventilation and air conditioning equipment. A constant supply of electricity is essential for safe ship and machinery operation, and therefore standby or additional capacity is necessary together with emergency supply equipment. Emergency equipment may take the form of an automatically starting emergency alternator or storage batteries may be used.

The complete range of electrical equipment will include generators, switch gear for control and distribution, motors and their associated starting equipment and emergency supply arrangements.

Alternating or direct current

Alternating current has now all but replaced direct current as the standard supply for all marine installations. The use of alternating current has a number of important advantages: for example, reduced first cost, less weight, less space required and a reduction in maintenance requirements. Direct current does, however, offer advantages in motor control using, for example, the Ward–Leonard system which provides a wide range of speed.

Machine rating

Motors and generators, both d.c. and a.c., are rated as Continuous Maximum Rated (CMR) machines. This means they can accept a considerable momentary overload and perhaps even a moderate overload for a longer duration.

Temperature affects the performance of all electrical equipment and also the useful life of the insulation and thus the equipment itself. The total temperature of an operating machine is a result of the ambient air temperature and the heating effect of current in the windings. Temperature rise is measured above this total temperature. Adequate ventilation of electrical equipment is therefore essential. Classification Societies have set requirements for the various classes of insulation. The usual classes for marine installations are E, B and F where particular insulation materials are specified and increasing temperature rises allowed in the order stated.

Enclosures

Depending upon the location, a motor or generator will have one of a number of possible types of enclosure. 'Drip-proof' is most common and provides protection from falling liquids or liquids being drawn in by ventilating air. A 'watertight enclosure' provides protection for immersion under a low head of water for up to one hour. 'Weatherproof', 'hose proof' and 'deck watertight' provide immersion protection for only one minute. 'Totally enclosed' can also be used or an arrangement providing ducted ventilation from outside the machinery space. A 'flameproof' enclosure is capable of withstanding an explosion of some particular flammable gas that may occur within it. It must also stop the transfer of flame, i.e. contain any fire or explosion.

Direct current generators

A current is produced when a single coil of wire is rotated in a magnetic field. When the current is collected using a ring which is split into two halves (a *commutator*), a direct or single direction current is produced. The current produced may be increased by the use of many turns of wire and additional magnetic fields.

With many coils connected to the commutator, sparking will occur as the current collecting brushes move across the insulated segments. Commutating poles or interpoles are used to reduce this sparking. They are in fact electromagnets having a polarity the same as the main pole which follows in the direction of rotation.

The magnetic field between the poles is produced by what are known as 'field coils'. These coils are excited or energised by the current produced in the machine. The soft iron core of the field coils retains some magnetism which enables a preliminary current generation to build up eventually to the full machine output. The field windings can be connected to the output current in a number of ways—shunt, series or compound. The compound wound arrangement is usual since it provides the best voltage characteristics.

The compound wound generator has two sets of field coils (Figure 14.1(a)). The shunt coil has many turns of fine wire and the series coil has a few turns of heavy wire. The shunt field produces full voltage on no-load which falls off as the load current increases. The series field creates an increase in voltage as the load increases. Properly combined or compounded the result is a fairly constant voltage over a range of load (Figure 14.1(b)).

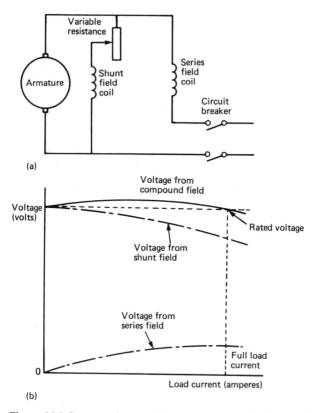

Figure 14.1 Compound wound d.c. generator, (a) field connection; (b) characteristic curves

Direct current distribution

The generated supply is provided to conductors known as 'bus-bars' which are located behind the main switchboard. The supply then passes through circuit breakers to auxiliaries directly or to section or distribution boards. A circuit breaker is an isolating switch. A section board is a grouping of electrical services fed from the main board. A distribution board feeds minor supplies such as lighting and may itself be fed from the main board or a section board. The distribution system is shown in Figure 14.2.

A two wire system is usual to provide a supply and return to each item of equipment. An earth lead would be the only electrical connection between any item of equipment and the ship's structure. With compound wound generators a third bus-bar would be introduced as the equalising connection between machines.

A fuse is a type of switch which isolates a circuit if an excessive current flows. To reconnect the circuit, after discovering the cause of the overload, the fuse must be rewired or replaced. The fuse is in effect a weak link in the circuit designed to break and protect equipment from damaging high currents. A semi-enclosed or rewirable fuse will have provision for a wire to be replaced after it has burnt out. The correct rating of fuse wire should be replaced within the holder to reinstate the circuit. A cartridge fuse has the wire enclosed within a ceramic body and it is not rewirable. A 'blown' cartridge fuse must be replaced by a new one. The cartridge fuse is to be preferred since the fusing current value is more reliable than for a rewirable type.

A circuit breaker is an isolating switch which also functions as a fuse. It has two designed ratings: one of the normal safe working current, the other the overload current. The breaker is closed against the action of a spring to make the circuit and supply the section board or auxiliary. A trip mechanism opens the breaker, a fast opening being ensured by the spring. When desired the breaker is tripped or opened manually. It will also open if the overload current rating is exceeded for a period of time.

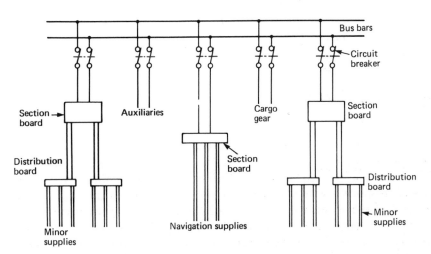

Figure 14.2 D.C. distribution system

A delay mechanism prevents the breaker opening for short-period overload currents. The circuit breaker opens or closes both supply and return leads in the circuit. Where a circuit breaker feeds the generator supply to the bus-bars a third 'make-or-break' arm will be provided for the equaliser connection.

Preferential tripping is a means of retaining essential electrical supplies. In the event that a generator cannot supply all the load then non-essential loads are disconnected by preferential trips. The intention is to reduce the generator load while ensuring essential equipment such as steering gear, navigation lights, etc., retains its electrical supply.

Various circuit faults can occur as a result of either a break in the conductor (cable) or a break in the insulation. An open-circuit fault results from a break in the conductor and no current flow will take place. A short-circuit fault is due to two breaks in the insulation on, for example, adjacent conductors. The two conductors are connected and a large current flow takes place. An earth fault occurs when a break in the insulation permits the conductor to touch an earthed metal enclosure (or the hull).

Earth faults are usually detected by the use of earth indicating lamps. Two lamps are used, each rated for the full system voltage, but connected in series across the system with the mid point earthed (Figure 14.3). If the system is correctly insulated then both lamps will glow at

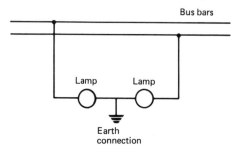

Figure 14.3 Earth lamp circuit

half brilliance. The lamps are placed close together to enable a comparison to be made. A direct earth in one pole will short circuit its lamp, causing the other to shine brightly. A slight insulation breakdown would produce a difference in bulb brightness between the two. Where an earth fault is detected the circuit breakers for each separate circuit must be opened in turn until the fault location is discovered. The particular section or distribution box would then have to have its circuits investigated one by one to locate the fault and enable its correction.

Direct current supply

The supply to a distribution system will usually come from two or more generators operating in parallel. Each generator must be provided with certain protective devices to ensure against reverse currents, low voltage or an overcurrent. There must also be ammeters and voltmeters in the circuits to enable paralleling to take place.

The circuit for two generators operating in parallel is shown in Figure 14.4. A triple-pole circuit breaker connects the supply to the bus-bars and also the equaliser bus-bar. The arrangement of the various protective trips can be seen, with excess current protection being provided in each pole. The reverse current trip prevents the generator operating as a motor if, for instance, the prime mover stopped.

The voltmeters and ammeters are provided in the generator supply circuits for paralleling purposes. A voltmeter is positioned across the bus-bars to indicate their voltage. Consider the situation where one

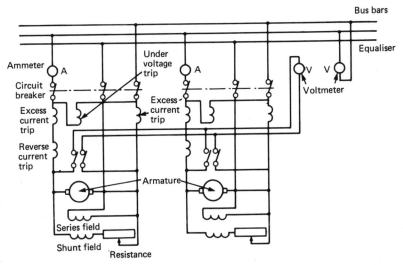

Figure 14.4 Protective trips for the parallel operation of two d.c. generators

generator is supplying the bus-bar system and a second generator is to be paralleled with it. The second machine is run up to speed and its field current adjusted until the two machines are at the same voltage. The circuit breaker connecting the second machine to the bus-bar can now be closed and the field current adjusted to enable the generator to take its share of the load. When the load is evenly shared the two machines can then be left to operate in parallel. The equalising connection will cater for any slight changes in load sharing that occur.

Alternating current generators

A coil of wire rotating in a magnetic field produces a current. The current can be brought out to two sliprings which are insulated from the shaft. Carbon bushes rest on these rings as they rotate and collect the current for use in an external circuit. Current collected in this way will be alternating, that is, changing in direction and rising and falling in value. To increase the current produced, additional sets of poles may be introduced.

The magnetic field is provided by electromagnets so arranged that adjacent poles have opposite polarity. These 'field coils', as they are called, are connected in series to an external source or the machine output.

If separate coils or conductors are used then several outputs can be obtained. Three outputs are usually arranged with a phase separation of 120°, to produce a three-phase supply. The supply phasing is shown in Figure 14.5. The three-phase system is more efficient in that for the same mechanical power a greater total electrical output is obtained. Each of the three outputs may be used in single-phase supplies or in conjunction for a three-phase supply. The separate supplies are connected in either star or delta formation (Figure 14.6). The star formation is most commonly used and requires four sliprings on the alternator. The three conductors are joined at a common slipring and also have their individual slipring. The central or neutral line is common to each phase. The delta arrangement has two phases joined at each of the three sliprings on the alternator. A single-phase supply can be taken from any two sliprings.

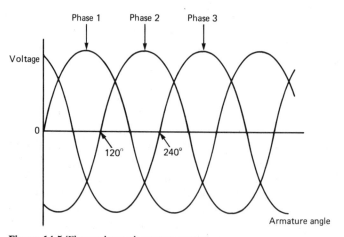

Figure 14.5 Three-phase alternator output

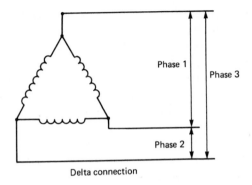

Delta connection

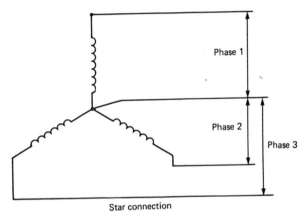

Star connection

Figure 14.6 Star and delta three-phase connections

So far, alternator construction has considered the armature rotating and the field coils stationary. The same electricity generating effect is produced if the reverse occurs, that is, the field coils rotate and the armature is stationary. This is in fact the arrangement adopted for large, heavy duty alternators.

The field current supply in older machines comes from a low-voltage direct current generator or exciter on the same shaft as the alternator. Modern machines, however, are either statically excited or of the high-speed brushless type. The exciter is required to operate to counter the effects of power factor for a given load. The power factor is a measure of the phase difference between voltage and current and is expressed as the cosine of the phase angle. With a purely resistance load the voltage and current are in phase, giving a power factor of one. The power consumed is therefore the product of voltage and current. Inductive or capacitive loads, combined with resistance loads, produce lagging or leading power factors which have a value less than one. The power consumed is the product of current, voltage and power factor. The alternating current generator supplying a load has a voltage drop resulting from the load. When the load has a lagging power factor this voltage drop is considerable. Therefore the exciter, in maintaining the alternator voltage, must vary with the load current and also the power factor. The speed change of the prime mover must also be taken into account.

Hand control of excitation is difficult so use is made of an automatic voltage regulator (AVR). The AVR consists basically of a circuit fed from the alternator output voltage which detects small changes in voltage and feeds a signal to an amplifier which changes the excitation to correct the voltage. Stabilising features are also incorporated in the circuits to avoid 'hunting' (constant voltage fluctuations) or overcorrecting. Various designs of AVR are in use which can be broadly divided into classes such as carbon pile types, magnetic amplifiers, electronic types, etc.

The statically excited alternator has a static excitation system instead of a d.c. exciter. This type of alternator will more readily accept the sudden loading by direct on-line starting of large squirrel cage motors. The static excitation system uses transformers and rectifiers to provide series and shunt components for the alternator field, that is, it is compounded. Brushes and sliprings are used to transfer the current to the field coils which are mounted on the rotor. The terminal voltage from the alternator thus gives the no-load voltage and the load current

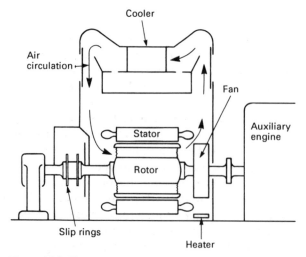

Figure 14.7 Alternator construction

provides the extra excitation to give a steady voltage under any load condition. The careful matching of components provides a system which functions as a self regulator of voltage. Certain practical electrical problems and the compensation necessary for speed variation require that a voltage regulator is also built into the system.

The brushless high speed alternator was also developed to eliminate d.c. exciters with their associated commutators and brushgear. The alternator and exciter rotors are on a common shaft, which also carries the rectifiers. The exciter output is fed to the rectifiers and then through conductors in the hollow shaft to the alternator field coils. An automatic voltage regulator is used with this type of alternator.

The construction of an alternator can be seen in Figure 14.7. The rotor houses the poles which provide the field current, and these are usually of the salient or projecting-pole type. Slip rings and a fan are also mounted on the rotor shaft, which is driven by the auxiliary engine. The stator core surrounds the rotor and supports the three separate phase windings. Heat is produced in the various windings and must be removed by cooling. The shaft fan drives air over a water-cooled heat exchanger. Electric heaters are used to prevent condensation on the windings when the alternator is not in use.

In addition to auxiliary-engine-driven alternators a ship may have a shaft-driven alternator. In this arrangement a drive is taken from the main engine or the propeller shaft and used to rotate the alternator. The various operating conditions of the engine will inevitably result in variations of the alternator driving speed. A hydraulic pump and gearbox arrangement may be used to provide a constant-speed drive, or the alternator output may be fed to a static frequency converter. In the static frequency converter the a.c. output is first rectified into a variable d.c. voltage and then inverted back into a three-phase a.c. voltage. A feedback system in the oscillator inverter produces a constant-output a.c. voltage and frequency.

Distribution system

An a.c. distribution system is provided from the main switchboard which is itself supplied by the alternators (Figure 14.8). The voltage at the switchboard is usually 440 volts, but on some large installations it may be as high as 3300 volts. Power is supplied through circuit breakers to larger auxiliaries at the high voltage. Smaller equipment may be supplied via fuses or miniature circuit breakers. Lower voltage supplies used, for instance, for lighting at 220 volts, are supplied by step down transformers in the distribution network.

The distribution system will be three-wire with insulated or earthed neutral. The insulated neutral has largely been favoured, but earthed neutral systems have occasionally been installed. The insulated neutral system can suffer from surges of high voltage as a result of switching or system faults which could damage machinery. Use of the earthed system could result in the loss of an essential service such as the steering gear as a result of an earth fault. An earth fault on the insulated system would not, however, break the supply and would be detected in the earth lamp display. Insulated systems have therefore been given preference since earth faults are a common occurrence on ships and a loss of supply in such situations cannot be accepted.

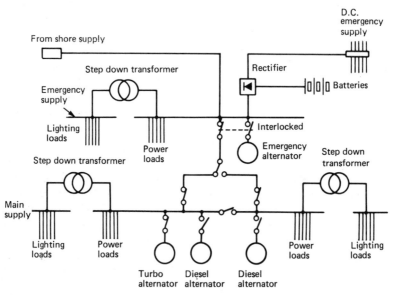

Figure 14.8 A.C. distribution system

In the distribution system there will be circuit breakers and fuses, as mentioned previously for d.c. distribution systems. Equipment for a.c. systems is smaller and lighter because of the higher voltage and therefore lower currents. Miniature circuit breakers are used for currents up to about 100 A and act as a fuse and a circuit breaker. The device will open on overload and also in the event of a short circuit. Unlike a fuse, the circuit can be quickly remade by simply closing the switch. A large version of this device is known as the 'moulded-case circuit breaker' and can handle currents in excess of 1000 A. Preferential tripping and earth fault indication will also be a part of the a.c. distribution system. These two items have been mentioned previously for d.c. distribution systems.

Alternating current supply

Three-phase alternators arranged for parallel operation require a considerable amount of instrumentation. This will include ammeters, wattmeter, voltmeter, frequency meter and a synchronising device. Most of these instruments will use transformers to reduce the actual values taken to the instrument. This also enables switching, for instance, between phases or an incoming machine and the bus-bars, so that one instrument can display one of a number of values. The wattmeter measures the power being used in a circuit, which, because of the power factor aspect of alternating current load, will be less than the product of the volts and amps. Reverse power protection is provided to alternators since reverse current protection cannot be used. Alternatively various trips may be provided in the event of prime mover failure to ensure that the alternator does not act as a motor.

The operation of paralleling two alternators requires the voltages to be equal and also in phase. The alternating current output of any machine is always changing, so for two machines to operate together their voltages must be changing at the same rate or frequency and be reaching their maximum (or any other value) together. They are then said to be 'in phase'. Use is nowadays made of a synchroscope when paralleling two a.c. machines. The synchroscope has two windings which are connected one to each side of the paralleling switch. A pointer is free to rotate and is moved by the magnetic effect of the two windings. When the two voltage supplies are in phase the pointer is stationary in the 12 o'clock position. If the pointer is rotating then a frequency difference exists and the dial is marked for clockwise rotation FAST and anti-clockwise rotation SLOW, the reference being to the incoming machine frequency.

To parallel an incoming machine to a running machine therefore it is necessary to ensure firstly that both voltages are equal. Voltmeters are provided for this purpose. Secondly the frequencies must be brought into phase. In practice the synchroscope usually moves slowly in the FAST direction and the paralleling switch is closed as the pointer reaches the 11 o'clock position. This results in the incoming machine immediately accepting a small amount of load.

A set of three lamps may also be provided to enable synchronising. The sequence method of lamp connection has a key lamp connected across one phase with the two other lamps cross connected over the other two phases. If the frequencies of the machines are different the lamps will brighten and darken in rotation, depending upon the incoming frequency being FAST or SLOW. The correct moment for synchronising is when the key lamp is dark and the other two are equally bright.

Direct current motors

When a current is supplied to a single coil of wire in a magnetic field a force is created which rotates the coil. This is a similar situation to the generation of current by a coil moving in a magnetic field. In fact generators and motors are almost interchangeable, depending upon which two of magnetic field, current and motion are provided. Additional coils of wire and more magnetic fields produce a more efficient motor. Interpoles are fitted to reduce sparking but now have opposite polarity to the next main pole in the direction of rotation. When rotating the armature acts as a generator and produces current in the reverse direction to the supply. This is known as back e.m.f. (electromotive force) and causes a voltage drop across the motor. This back e.m.f. controls the power used by the motor but is not present as the motor is started. As a result, to avoid high starting currents special control circuits or starters are used.

The behaviour of the d.c. motor on load is influenced by the voltage drop across the armature, the magnetic field produced between the poles and the load or torque on the motor. Some of these factors are interdependent. For example, the voltage drop across the armature depends upon the back e.m.f. which depends upon the speed of the motor and the strength of the magnetic field. Shunt, series and compound windings are used to obtain different motor characteristics by varying the above factors.

The shunt wound motor has field windings connected in parallel with the armature windings (Figure 14.9). Thus when the motor is operating with a fixed load at constant speed all other factors are constant. An increase in load will cause a drop in speed and therefore a reduction in back e.m.f. A greater current will then flow in the armature windings and the motor power consumption will rise: the magnetic field will be unaffected since it is connected in parallel. Speed reduction is, in

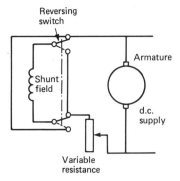

Figure 14.9 Shunt wound d.c. motor

practice, very small, which makes the shunt motor an ideal choice for constant-speed variable-load duties.

The series motor has field windings connected in series with the armature windings (Figure 14.10). With this arrangement an increase in load will cause a reduction in speed and a fall in back e.m.f. The increased load current will, however, now increase the magnetic field and therefore the back e.m.f. The motor will finally stabilise at some reduced value of speed. The series motor speed therefore changes considerably with load.

Control of d.c. motors is quite straightforward. The shunt wound motor has a variable resistance in the field circuit, as shown in Figure 14.9. This permits variation of the current in the field coils and also the back e.m.f., giving a range of constant speeds. To reverse the motor the field current supply is reversed, as shown in Figure 14.9.

One method of speed control for a series wound motor has a variable resistance in parallel with the field coils. Reverse operation is again achieved by reversing the field current supply as shown in Figure 14.10.

In operation the shunt wound motor runs at constant speed regardless of load. The series motor runs at a speed determined by the load, the greater the load the slower the speed. Compounding—the use of shunt and series field windings—provides a combination of these characteristics. Starting torque is also important. For a series wound motor the starting torque is high and it reduces as the load increases. This makes the series motor useful for winch and crane applications. It should be noted that a series motor if started on no-load has an infinite speed. Some small amount of compounding is usual to avoid this dangerous occurrence. The shunt wound motor is used where constant speed is required regardless of load; for instance, with fans or pumps.

The starting of a d.c. motor requires a circuit arrangement to limit armature current. This is achieved by the use of a starter (Figure 14.11). A number of resistances are provided in the armature and progressively removed as the motor speeds up and back e.m.f. is developed. An arm, as part of the armature circuit, moves over resistance contacts such that a number of resistances are first put into the armature circuit and then

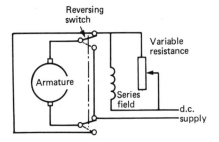

Figure 14.10 Series wound d.c. motor

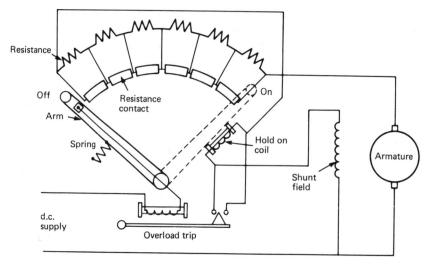

Figure 14.11 D.C. motor starter

progressively removed. The arm must be moved slowly to enable the motor speed and thus the back e.m.f. to build up. At the final contact no resistance is in the armature circuit. A 'hold on' or 'no volts' coil holds the starter arm in place while there is current in the armature circuit. If a loss of supply occurs the arm will be released and returned to the 'off' position by a spring. The motor must then be started again in the normal way. An overload trip is also provided which prevents excess current by shorting out the 'hold on' coil and releasing the starter arm. The overload coil has a soft iron core which, when magnetised sufficiently by an excess current, attracts the trip bar which shorts out the hold on coil. This type of starter is known as a 'face plate'; other types make use of contacts without the starting handle but introduce resistance into the armature circuit in much the same way.

Alternating current motors

Supplying alternating current to a coil which is free to rotate in a magnetic field will not produce a motor effect since the current is constantly changing direction. Use is therefore made in an induction or squirrel cage motor of a rotating magnetic field produced by three separately phased windings in the stator. The rotor has a series of copper conductors along its axis which are joined by rings at the ends to form a cage. When the motor is started the rotating magnetic field induces an e.m.f. in the cage and thus a current flow. The current-carrying conductor in a magnetic field produces the motor effect which turns the rotor. The motor speed builds up to a value just less than the speed of rotation of the magnetic field.

The motor speed depends upon the e.m.f. induced in the rotor and this depends upon the difference in speed between the conductors and the magnetic field. If the load is increased the rotor slows down slightly, causing an increase in induced e.m.f. and thus a greater torque to deal with the increased load. The motor is almost constant speed over all values of load. It will start against about two times full load torque but draws a starting current of about six times the normal full load current. The starting current can be reduced by having a double cage arrangement on the rotor. Two separated cages are provided, one below the other in the rotor. When starting, the outer high-resistance cage carries almost all the rotor current. As the motor accelerates the low-resistance inner winding takes more and more of the current until it carries the majority.

A number of different fixed speeds are possible by pole changing. The speed of an induction motor is proportional to frequency divided by the numbers of pairs of poles. If therefore a switch is provided which can alter the numbers of pairs of poles, then various fixed speeds are possible. The number of poles affects the starting characteristics such that the more poles the less the starting torque to full load torque ratio.

Only the induction type of a.c. motor has been described, since it is almost exclusively used in maritime work. Synchronous motors are another type which have been used for electrical propulsion systems but not auxiliary drives.

A number of different arrangements can be used for starting an induction motor. These include direct on-line, star delta, auto transformer and stator resistance. Direct on-line starting is usual where the distribution system can accept the starting current. Where a slow moving high inertia load is involved the starting time must be considered because of the heating effect of the starting current. The star delta starter connects the stator windings first in star and when running changes over to delta. The star connection results in about half of the line voltage being applied to each phase with therefore a reduction in starting current. The starting torque is also reduced to about one-third of its direct on line value. A rapid change-over to delta is required at about 75% of full load speed when the motor will draw about three-and-a-half times its full load current. The auto transformer starter is used only for large motors. It uses tappings from a transformer to provide, for example, 40%, 60% and 75% of normal voltage (Figure 14.12). The motor is started on one of the tappings and then quickly switched to full voltage at about 75% full speed. The tapping chosen will depend upon the starting torque required with a 60% tapping giving

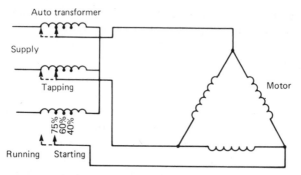

Figure 14.12 Squirrel cage induction motor starting

about 70% of full load torque. A smaller percentage tapping will give a smaller starting torque and vice-versa. The stator resistance starter has a resistance in the stator circuit when the motor is started. An adjustable timing device operates to short circuit this resistance when the motor has reached a particular speed.

Modern electronic techniques enable a.c. induction motors to be used in speed-control systems. The ship's supply, which may not be as stable in voltage or frequency as that ashore, is first rectified to provide a d.c. supply. This is then used as the power supply of an oscillator using high-power electronic devices. These may be thyristors (for powers up to 1.5 MW or more) or transistors (for powers up to a few tens of kilowatts). The high-power oscillator output is controlled in frequency and voltage by a feedback system. The motor speed is varied by changing the oscillator output frequency. The motor current necessary to obtain the desired torque (at small angles of slip) is normally obtained by maintaining the voltage almost proportional to frequency.

Certain protective devices are fitted in the motor circuit to protect against faults such as single phasing, overload or undervoltage. Single phasing occurs when one phase in a three-phase circuit becomes open circuited. The result is excessive currents in all the windings with, in the case of a delta connected stator running at full load, one winding taking three times its normal load current. A machine which is running when single phasing occurs will continue to run but with an unbalanced distribution of current. An overload protection device may not trip if the motor is running at less than full load. One method of single phasing protection utilises a temperature-sensitive device which isolates the machine from the supply at some particular winding temperature. Overload protection devices are also fitted and may be separate or combined with the single phase protection device. They must have a time delay fitted so that operation does not occur during the high starting current period. An undervoltage or 'no volts' protective device ensures that the motor is properly started after a supply failure.

Maintenance

With all types of electrical equipment cleanliness is essential for good operation. Electrical connections must be sound and any signs of sparking should be investigated. Parts subject to wear must be examined and replaced when necessary. The danger from a.c. equipment in terms of electric shocks is far greater than for similar d.c. voltages. Also a.c. equipment often operates at very high voltages. Care must therefore be taken to ensure isolation of equipment before any inspections or maintenance is undertaken.

The accumulation of dirt on electrical equipment will result in insulation breakdown and leakage currents, possibly even an earth fault. Moisture or oil deposits will likewise affect insulation resistance. Regular insulation resistance measurement and the compiling of records will indicate the equipment requiring attention. Ventilation passages or ducts may become blocked, with resultant lack of cooling and overheating. Oil deposits from a direct-coupled diesel engine driving an open generator (usually d.c.) can damage windings and should therefore be removed if found. Totally enclosed machines should be periodically opened for inspection and cleaning since carbon dust will remain inside the machine and deposit on the surfaces.

Brushgear should be inspected to ensure adequate brush pressure and the springs adjusted if necessary. New brushes should be 'bedded in' to the commutator or slipring shape with fine glass paper. Sparking at the commutator will indicate poor commutation. This may require polishing of a roughened commutator surface. The mica insulation between commutator segments may require undercutting if it protrudes, or simply cleaning if deposits have built up.

Control equipment, such as starters, will require attention to contacts which may be worn or pitted as a result of arcing. Contactors usually have a moving or wiping action as they come together. This helps clean the surfaces to provide good electrical contact, and also the arc produced during closing and opening is not at the finally closed position. The contactor contact surfaces of frequently used equipment should therefore be subject to regular inspections.

Batteries

The battery is a convenient means of storing electricity. It is used on many ships as an instantly available emergency supply. It may also be used on a regular basis to provide a low-voltage d.c. supply to certain equipment. To provide these services the appropriate size and type of battery must be used and should be regularly serviced. Two main types of battery are used on board ship: the lead–acid and the alkaline type, together with various circuits and control gear.

Lead–acid battery

The lead–acid battery is made up of a series of cells. One cell consists of a lead peroxide positive plate and a lead negative plate both immersed in a dilute sulphuric acid solution. The sulphuric acid is known as the 'electrolyte'. A wire joining these two plates will have a potential or voltage developed across it and a current will flow. This voltage is about 2.2 V initially with a steady value of about 2 V. A grouping of six separate cells connected in series will give a 12 V battery. The word 'accumulator' is sometimes used instead of battery.

Actual construction uses interleaved plates in the cell in order to produce a compact arrangement with a greater capacity. The complete battery is usually surrounded by a heavy-duty plastic, hard rubber or bitumen case.

In the charged condition the battery contains lead, lead peroxide and sulphuric acid. During discharge, i.e. the providing of electrical power, some of the lead peroxide and the lead will change to lead sulphate and water. The sulphuric acid is weakened by this reaction and its specific gravity falls.

When the battery is charged, i.e. electrical power is put into it, the reactions reverse to return the plates to their former material and the water produced breaks down into hydrogen gas which bubbles out.

Alkaline battery

The basic cell of the alkaline battery consists of a nickel hydroxide positive plate and a cadmium and iron negative plate immersed in a solution of potassium hydroxide. The cell voltage is about 1.4 V. A grouping of five cells is usual to give about seven volts.

An interleaved construction is again used and each cell is within a steel casing. This casing is electrically 'live' being in contact with the electrolyte and possibly one set of plates. A battery consists of a group of cells mounted in hardwood crates with space between each. The cells are connected in series to give the battery voltage.

In the charged condition the positive plate is nickel hydroxide and the negative plate cadmium. During discharge oxygen is transferred from one plate to the other without affecting the specific gravity of the potassium hydroxide solution. The negative plate becomes cadmium oxide and the positive plate is less oxidised nickel hydroxide. Charging the battery returns the oxygen to the positive plate.

Battery selection

The choice between the lead–acid or alkaline type of battery will be based upon their respective advantages and disadvantages.

The lead–acid battery uses fewer cells to reach a particular voltage. It is reasonably priced but has a limited life. It does, however, discharge on open circuit and requires regular attention and charging to keep it in a fully charged condition. If left in a discharged condition for any period of time a lead–acid battery may be ruined.

The alkaline battery retains its charge on open circuit and even if discharged it can be left for long periods without any adverse effect. Although more expensive it will last much longer and requires less attention. Also a greater number of cells are required for a particular voltage because of the smaller nominal value per cell.

Both types of battery are widely used at sea for the same basic duties.

Operating characteristics

When operating in a circuit a battery provides current and voltage and is itself discharging. Depending upon the capacity, it will provide current and voltage for a short or a long time. The capacity is measured in ampere hours, i.e. the number of hours a particular current can be supplied. Thus a 20 ampere-hour capacity battery can supply 2 A for 10 hours or 1 A for 20 hours. This is a reasonable assumption for small currents. The ampere-hour capacity does depend upon the rate of discharge and therefore for currents above about 5 A, a rate of discharge is also quoted.

Having been 'discharged' by delivering electrical power a battery must then be 'charged' by receiving electrical power. To charge the battery an amount of electrical power must be provided in the order of the capacity. Some energy loss occurs due to heating and therefore slightly more than the capacity in terms of electrical power must be provided. By charging with a low current value the heating losses can be kept to a minimum.

The different methods of charging include *constant current, constant voltage* and *trickle charge*. With constant current charging the series resistance is reduced in order to increase the charging voltage. This may be achieved manually or automatically. The constant voltage system results in a high value of current which gradually falls as the battery charges. The circuit resistance prevents the initial current from being too high. Trickle charging is used to keep a battery in peak condition—a very low current is continuously passed through the battery and keeps it fully charged.

Maintenance

To be available when required batteries must be maintained in a fully charged condition. Where lead–acid batteries are used this can be achieved by a constant trickle charge. Otherwise, for both types of battery, a regular charge-up is necessary.

A measure of the state of charge can be obtained by using a hydrometer. This is a device for measuring the specific gravity of a liquid. A syringe-type hydrometer is shown in Figure 14.13. A sample of electrolyte is taken from each cell in turn and its specific gravity is measured by reading the float level. All specific gravity values for the individual cells in a battery should read much the same. The specific gravity reading can be related to the state of charge of the battery. The specific gravity reading must be corrected for the temperature of the electrolyte. The value for a fully charged lead–acid battery is 1.280 at

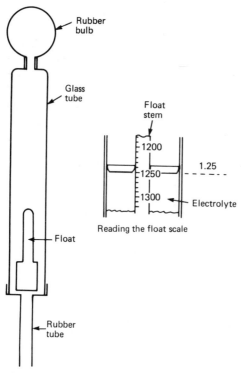

Figure 14.13 Syringe-type hydrometer

15°C. For an alkaline battery the specific gravity does not alter much during charge and discharge but gradually falls over a long period: when a value of 1.160 is reached it should be replaced.

The electrolyte level should be maintained just above the top of the plates. Any liquid loss due to evaporation or chemical action should be replaced with distilled water. Only in an emergency should other water be used. It is not usual to add electrolyte to batteries.

A battery must be kept clean and dry. If dirt deposits build up or spilt electrolyte remains on the casing, stray currents may flow and discharge the battery. Corrosion of the casing could also occur. The battery terminals should be kept clean and smeared with a petroleum jelly. The small vents in the cell caps should be clear at all times.

Cell voltage readings are useful if taken while the battery is discharging. All cells should give about the same voltage reading. This test method is of particular value with alkaline batteries, where specific gravity readings for the electrolyte do not indicate the state of charge.

Ward–Leonard speed control system

As a very flexible, reliable means of motor speed control the Ward–Leonard system is unmatched.

The system is made up of a driving motor which runs at almost constant speed and powers a d.c. generator (Figure 14.14). The generator output is fed to a d.c. motor. By varying the generator field current its output voltage will change. The speed of the controlled motor can thus be varied smoothly from zero to full speed. Since control

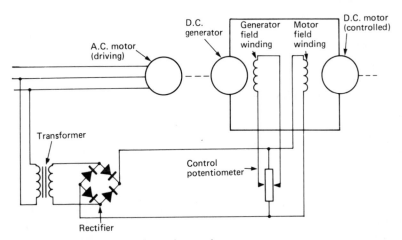

Figure 14.14 Ward–Leonard speed control

is achieved through the generator shunt field current, the control equipment required is only for small current values. A potentiometer or rheostat in the generator field circuit enables the variation of output voltage from zero to the full value and also in either direction. The controlled motor has a constant excitation: its speed and direction are thus determined by the generator output.

Depending upon the particular duties of the controlled motor, series windings may be incorporated in the field of the motor and also the generator. This may result in additional switching to reverse the controlled motor depending upon the compounding arrangements.

The driving motor or prime motor for the Ward–Leonard system can be a d.c. motor, an a.c. motor, a diesel engine, etc. Any form of constant or almost constant speed drive can be used, since its function is only to drive the generator.

Emergency generator supply

In the event of a main generating system failure an emergency supply of electricity is required for essential services. This can be supplied by batteries, but most merchant ships have an emergency generator. The unit is diesel driven and located outside of the machinery space (see Chapter 10, Emergency equipment).

The emergency generator must be rated to provide power for the driving motors of the emergency bilge pump, fire pumps, steering gear, watertight doors and possibly fire fighting equipment. Emergency lighting for occupied areas, navigation lights, communications systems and alarm systems must also be supplied. Where electrical control devices are used in the operation of main machinery, these too may require a supply from the emergency generator.

A switchboard in the emergency generator room supplies these various loads (Figure 14.8). It is not usual for an emergency generator to require paralleling, so no equipment is provided for this purpose. Automatic start up of the emergency generator at a low voltage value is usual on modern installations.

Navigation lights

The supply to the navigation lights circuit must be maintained under all circumstances and special provisions are therefore made.

To avoid any possibility of accidental open circuits the distribution board for the navigation lights supplies no other circuit. A changeover switch provides an alternative source of supply should the main supply fail. If the navigation lights fail, a visual or audible indication must be given.

A navigation lights circuit is shown in Figure 14.15. Two sources of supply are available from the changeover switch. A double pole switch connects the supply to each light circuit, with a fuse in each line. A relay in the circuit will operate the buzzer if an open circuit occurs, since the relay will de-energise and the trip bar will complete the buzzer circuit. A resistance in series with the indicating lamp will ensure the navigation lights operate even if the indicating lamp fails. A main supply failure will result in all the indicating lamps extinguishing but the buzzer will not sound. The changeover switch will then have to be moved to the alternative supply.

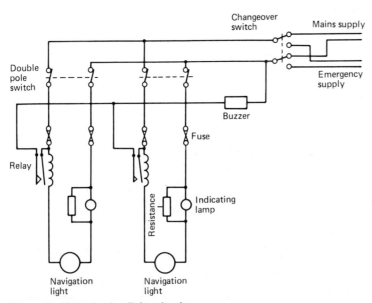

Figure 14.15 Navigation lights circuit

Insulation resistance measurement

Good insulation resistance is essential to the correct operation of electrical equipment. A means must be available therefore to measure insulation resistance. Readings taken regularly will give an indication as to when and where corrective action, maintenance, servicing, etc., is required.

Insulation resistance may be measured between a conductor and earth or between conductors. Dirt or other deposits on surfaces can reduce

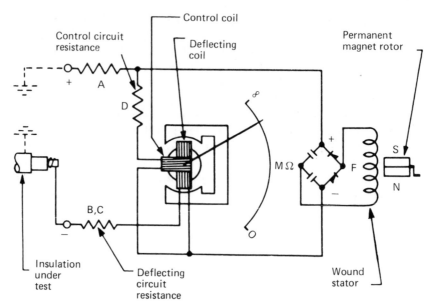

Figure 14.16 Insulation tester

insulation resistance and cause a leakage current or 'tracking' to occur. Equipment must therefore be kept clean in order to ensure high values, in megohms, of insulation resistance.

Insulation is classified in relation to the maximum temperature at which it is safe for the equipment or cables to operate. Classes A (55°C), E (70°C) and B (80°C) are used for marine equipment.

One instrument used for insulation testing is shown in Figure 14.16. Its trade name is 'Megger Tester'. A permanent magnet provides a magnetic field for a pivoted core which is wound with two coils. A needle or pointer is pivoted at the centre of rotation of the coils and moves when they do. The two coils are wound at right angles to each other and connected in such a way that one measures voltage and the other measures current. The needle deflection is a result of the opposing effects of the two coils which gives a reading of insulation resistance. A hand driven generator provides a test voltage to operate the instrument. Test probes are used to measure the resistance at the desired points.

Electrical hazards

The resistance of the human body is quite high only when the skin is dry. The danger of electric shock is therefore much greater for persons working in a hot, humid atmosphere since this leads to wetness from body perspiration. Fatal shocks have occurred at as low as 60 V and all circuits must be considered dangerous.

All electrical equipment should be isolated before any work is done on it. The circuit should then be tested to ensure that it is dead. Working near to live equipment should be avoided if at all possible. Tools with insulated handles should be used to minimise risks.

The treatment of anyone suffering from severe electric shock must be rapid if it is to be effective. First they must be removed from contact with the circuit by isolating it or using a non-conducting material to drag them away. Electric shock results in a stopping of the heart and every effort must be made to get it going again. Apply any accepted means of artificial respiration to bring about revival.

Chapter 15
Instrumentation and control

All machinery must operate within certain desired parameters. Instrumentation enables the parameters—pressure, temperature, and so on—to be measured or displayed against a scale. A means of control is also required in order to change or alter the displayed readings to meet particular requirements. Control must be manual, the opening or closing of a valve, or automatic, where a change in the system parameter results in actions which return the value to that desired without human involvement. The various display devices used for measurement of system parameters will first be examined and then the theory and application of automatic control.

Pressure measurement

The measurement of pressure may take place from one of two possible datums, depending upon the type of instrument used. *Absolute pressure* is a total measurement using zero pressure as datum. *Gauge pressure* is a measurement above the atmospheric pressure which is used as a datum. To express gauge pressure as an absolute value it is therefore necessary to add the atmospheric pressure.

Manometer

A U-tube manometer is shown in Figure 15.1. One end is connected to the pressure source; the other is open to atmosphere. The liquid in the tube may be water or mercury and it will be positioned as shown. The excess of pressure above atmospheric wil be shown as the difference in liquid levels; this instrument therefore measures *gauge pressure*. It is usually used for low value pressure readings such as air pressures. Where two different system pressures are applied, this instrument will measure *differential pressure*.

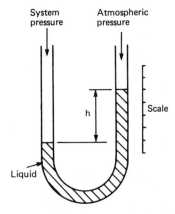

System pressure

Atmospheric pressure

Scale

h

Liquid

h = system pressure (gauge value)   **Figure 15.1** U-tube manometer

Barometer

The *mercury barometer* is a straight tube type of manometer. A glass capillary tube is sealed at one end, filled with mercury and then inverted in a small bath of mercury (Figure 15.2). Almost vacuum conditions exist above the column of mercury, which is supported by atmospheric

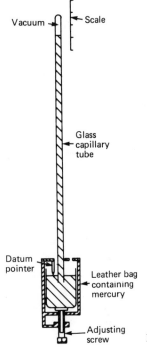

Vacuum → ⌐← Scale

Glass
← capillary
tube

Datum
pointer

Leather bag
← containing
mercury

Adjusting
screw

Figure 15.2 Mercury barometer

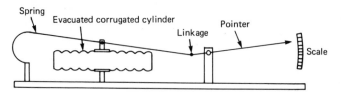

Figure 15.3 Aneroid barometer

pressure acting on the mercury in the container. An absolute reading of atmospheric pressure is thus given.

The *aneroid barometer* uses an evacuated corrugated cylinder to detect changes in atmospheric pressure (Figure 15.3). The cylinder centre tends to collapse as atmospheric pressure increases or is lifted by the spring as atmospheric pressure falls. A series of linkages transfers the movement to a pointer moving over a scale.

Bourdon tube

This is probably the most commonly used gauge pressure measuring instrument and is shown in Figure 15.4. It is made up of an elliptical

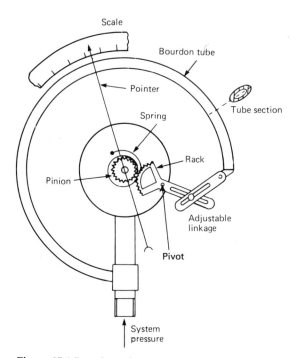

Figure 15.4 Bourdon tube pressure gauge

section tube formed into a C-shape and sealed at one end. The sealed end, which is free to move, has a linkage arrangement which will move a pointer over a scale. The applied pressure acts within the tube entering through the open end, which is fixed in place. The pressure within the tube causes it to change in cross section and attempt to straighten out with a resultant movement of the free end, which registers as a needle movement on the scale. Other arrangements of the tube in a helical or spiral form are sometimes used, with the operating principle being the same.

While the reference or zero value is usually atmospheric, to give gauge pressure readings, this gauge can be used to read vacuum pressure values.

Other devices

Diaphragms or bellows may be used for measuring gauge or differential pressures. Typical arrangements are shown in Figure 15.5. Movement of the diaphragm or bellows is transferred by a linkage to a needle or pointer display.

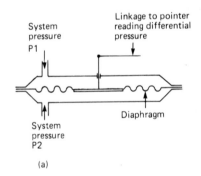

(a)

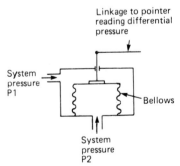

(b)

Figure 15.5 (a) Diaphragm pressure gauge; (b) bellows pressure gauge

The piezoelectric pressure transducer is a crystal which, under pressure, produces an electric current which varies with the pressure. This current is then provided to a unit which displays it as a pressure value.

Temperature measurement

Temperature measurement by instruments will give a value in degrees Celsius (°C). This scale of measurement is normally used for all readings and temperature values required except when dealing with theoretical calculations involving the gas laws, when absolute values are required (see Appendix).

Liquid-in-glass thermometer

Various liquids are used in this type of instrument, depending upon the temperature range, e.g. mercury −35°C to +350°C, alcohol −80°C to +70°C. An increase in temperature causes the liquid to rise up the narrow glass stem and the reading is taken from a scale on the glass (Figure 15.6). High-temperature-measuring mercury liquid thermometers will have the space above the mercury filled with nitrogen under pressure.

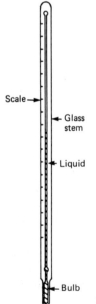

Figure 15.6 Liquid-in-glass thermometer

Liquid-in-metal thermometer

The use of a metal bulb and capillary bourdon tube filled with liquid offers advantages of robustness and a wide temperature range. The use of mercury, for instance, provides a range from −39°C to +650°C. The bourdon tube may be spiral or helical and on increasing temperature it tends to straighten. The free end movement is transmitted through linkages to a pointer moving over a scale.

Bimetallic strip thermometers

A bimetallic strip is made up of two different metals firmly bonded together. When a temperature change occurs different amounts of expansion occur in the two metals, causing a bending or twisting of the strip. A helical coil of bimetallic material with one end fixed is used in one form of thermometer (Figure 15.7). The coiling or uncoiling of the

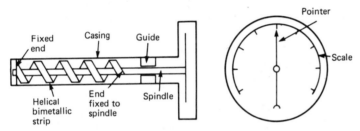

Figure 15.7 Bimetallic strip thermometer

helix with temperature change will cause movement of a pointer fitted to the free end of the bimetallic strip. The choice of metals for the strip will determine the range, which can be from −30°C to +550°C.

Thermocouple

The thermocouple is a type of electrical thermometer. When two different metals are joined to form a closed circuit and exposed to different temperatures at their junction a current will flow which can be used to measure temperature. The arrangement used is shown in Figure 15.8, where extra wires or compensating leads are introduced to complete the circuit and include the indicator. As long as the two ends A and B are at the same temperature the thermoelectric effect is not influenced. The appropriate choice of metals will enable temperature ranges from −200°C to +1400°C.

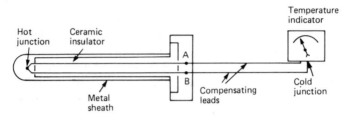

Figure 15.8 Thermocouple

Radiation pyrometer

A pyrometer is generally considered to be a high-temperature measuring thermometer. In the optical, or disappearing filament, type shown in Figure 15.9, radiation from the heat source is directed into the unit. The current through a heated filament lamp is adjusted until, when viewed through the telescope, it seems to disappear. The radiation from the lamp and from the heat source are therefore the same. The current through the lamp is a measure of the temperature of the heat source,

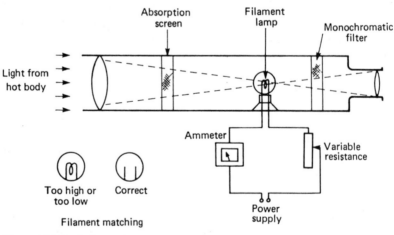

Figure 15.9 Optical pyrometer

and the ammeter is calibrated in units of temperature. The absorption screen is used to absorb some of the radiant energy from the heat source and thus extend the measuring range of the instrument. The monochromatic filter produces single-colour, usually red, light to simplify filament radiation matching.

Thermistor

This is a type of electrical thermometer which uses resistance change to measure temperature. The thermistor is a semi-conducting material made up of finely divided copper to which is added cobalt, nickel and manganese oxides. The mixture is formed under pressure into various shapes, such as beads or rods, depending upon the application. They are usually glass coated or placed under a thin metal cap.

A change in temperature causes a fall in the thermistor resistance which can be measured in an electric circuit and a reading relating to temperature can be given. Their small size and high sensitivity are particular advantages. A range of measurement from −250°C to +1500°C is possible.

Level measurement

Float operated

A float is usually a hollow ball or cylinder whose movement as the liquid surface rises or falls is transmitted to an indicator. A chain or wire usually provides the linkage to the indicator. Float switches may be used for high or low indication, pump starting, etc., where electrical contacts are made or broken, depending upon the liquid level.

Sight or gauge glasses

Various types of sightglass are used to display liquid level in storage tanks. The simple boiler gauge glass referred to in Chapter 4 is typical of such devices.

Pneumatic gauge

This is a device which uses a mercury manometer in conjunction with a hemispherical bell and piping to measure tank level. The arrangement is shown in Figure 15.10. A hemispherical bell is fitted near the bottom of the tank and connected by small bore piping to the mercury manometer. A selector cock enables one manometer to be connected to a number of tanks, usually a pair. A three-way cock is fitted to air, gauge and vent positions. With the cock at the 'air' position the system is filled with compressed air. The cock is then turned to 'gauge' when the tank contents will further pressurise the air in the system and a reading will be given on the manometer which corresponds to the liquid level. The cock is turned to 'vent' after the reading has been taken.

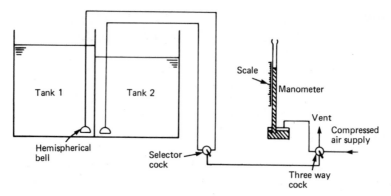

Figure 15.10 Pneumatic gauge

Flow measurement

Flow measurement can be quantity measurement, where the amount of liquid which has passed in a particular time is given, or a flow velocity which, when multiplied by the pipe area, will give a rate of flow.

Quantity measurement

A rotating pair of intermeshing vanes may be used which are physically displaced by the volume of liquid passing through (Figure 15.11(a)). The

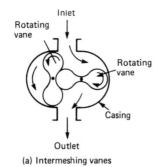

(a) Intermeshing vanes

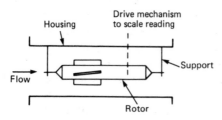

(b) rotating element

Figure 15.11 Flow quantity measurement

number of rotations will give a measure of the total quantity of liquid that has passed. The rotation transfer may be by mechanical means, such as gear wheels, or the use of a magnetic coupling.

Another method is the use of a rotating element which is set in motion by the passing liquid (Figure 15.11(b)). A drive mechanism results in a reading on a scale of total quantity. The drive mechanism may be mechanical, using gear wheels or electrical where the rotating element contains magnets which generate a current in a pick-up coil outside the pipe.

Flow velocity measurement

The venturi tube

This consists of a conical convergent entry tube, a cylindrical centre tube and a conical divergent outlet. The arrangement is shown in Figure 15.12. Pressure tappings led to a manometer will give a difference in

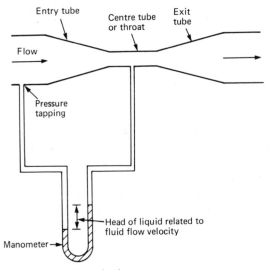

Figure 15.12 Venturi tube

head related to the fluid flow velocity. The operating principle is one of pressure conversion to velocity which occurs in the venturi tube and results in a lower pressure in the cylindrical centre tube.

The orifice plate

This consists of a plate with an axial hole placed in the path of the liquid. The hole edge is square facing the incoming liquid and bevelled on the

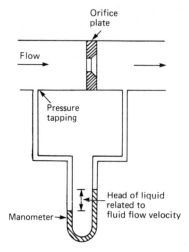

Figure 15.13 Orifice plate

outlet side (Figure 15.13). Pressure tappings before and after the orifice plate will give a difference in head on a manometer which can be related to liquid flow velocity.

Other variables

Moving coil meter

Electrical measurements of current or voltage are usually made by a moving coil meter. The meter construction is the same for each but its arrangement in the circuit is different.

A moving coil meter consists of a coil wound on a soft iron cylinder which is pivoted and free to rotate (Figure 15.14). Two hair springs are used, one above and one below, to provide a restraining force and also to

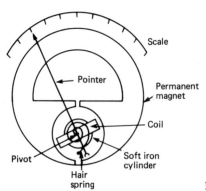

Figure 15.14 Moving coil meter

conduct the current to the coil. The moving coil assembly is surrounded by a permanent magnet which produces a radial magnetic field. Current passed through the coil will result in a force which moves the coil against the spring force to a position which, by a pointer on a scale, will read current or voltage.

The instrument is directional and must therefore be correctly connected in the circuit. As a result of the directional nature of alternating current it cannot be measured directly with this instrument, but the use of a rectifying circuit will overcome this problem.

Tachometers

A number of speed measuring devices are in use utilising either mechanical or electrical principles in their operation.

Mechanical

A simple portable device uses the governor principle to obtain a measurement of speed.

Two masses are fixed on leaf springs which are fastened to the driven shaft at one end and a sliding collar at the other (Figure 15.15). The

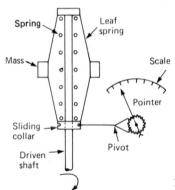

Figure 15.15 Mechanical tachometer

sliding collar, through a link mechanism, moves a pointer over a scale. As the driven shaft increases in speed the weights move out under centrifugal force, causing an axial movement of the sliding collar. This in turn moves the pointer to give a reading of speed.

Electrical

The drag cup generator device uses an aluminium cup which is rotated in a laminated iron electromagnet stator (Figure 15.16). The stator has

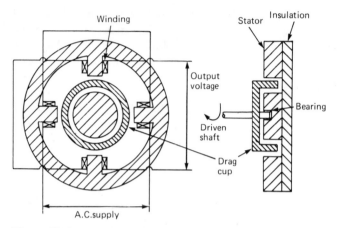

Figure 15.16 Drag cup generator-type tachometer

two separate windings at right angles to each other. An a.c. supply is provided to one winding and eddy currents are set up in the rotating aluminium cup. This results in an induced e.m.f. in the other stator winding which is proportional to the speed of rotation. The output voltage is measured on a voltmeter calibrated to read in units of speed.

Tachogenerators provide a voltage value which is proportional to the speed and may be a.c. or d.c. instruments. The d.c. tachogenerator is a small d.c. generator with a permanent field. The output voltage is proportional to speed and may be measured on a voltmeter calibrated in units of speed. The a.c. tachogenerator is a small brushless alternator with a rotating multi-pole permanent magnet. The output voltage is again measured by a voltmeter although the varying frequency will affect the accuracy of this instrument.

Various pick-up devices can be used in conjunction with a digital counter to give a direct reading of speed. An inductive pick-up tachometer is shown in Figure 15.17(a). As the individual teeth pass the coil they induce an e.m.f. pulse which is appropriately modified and then fed to a digital counter. A capacitive pick-up tachometer is shown in Figure 15.17(b). As the rotating vane passes between the plates a capacitance change occurs in the form of a pulse. This is modified and then fed to the digital counter.

Torsionmeters

The measurement of torsion is usually made by electrical means. The twisting or torsion of a rotating shaft can be measured in a number of different ways to give a value of applied torque. Shaft power can then be calculated by multiplying the torque by the rotational speed of the shaft.

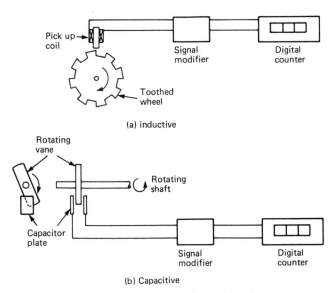

(a) inductive

(b) Capacitive

Figure 15.17 Pick-up tachometers, (a) inductive; (b) capacitive

Strain gauge torsionmeter

With this device four strain gauges are mounted onto the shaft, as shown in Figure 15.18. The twisting of the shaft as a result of an applied torque results in a change in resistance of the strain gauge system or bridge. Brushes and sliprings are used to take off the electrical connections and complete the circuit, as shown. More recently use has been made of the resistance change converted to a frequency change. A frequency converter attached to the shaft is used for this purpose: this frequency

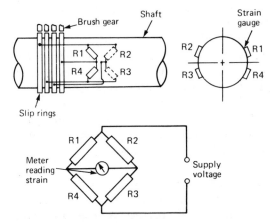

Figure 15.18 Strain gauge torsionmeter

signal is then transmitted without contact to a digital frequency receiver. When a torque is applied to the shaft, readings of strain and hence torque can be made.

Differential transformer torsionmeter

Two castings are used to provide a magnetic circuit with a variable air gap. The two are clamped to the shaft, as shown in Figure 15.19, and joined to each other by thin steel strips. The joining strips will transmit tension but offer no resistance to rotational movement of the two

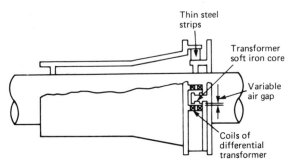

Figure 15.19 Differential transformer torsionmeter

castings with respect to each other. A differential transformer is fitted between the two castings, the two coils being wound on one casting and the iron core being part of the other. Another differential transformer is fitted in the indicating circuit, its air gap being adjusted by a micrometer screw. The primary coils of the two transformers are joined in series and energised by an a.c. supply. The secondary coils are connected so that the induced e.m.f.s are opposed and when one transformer has an air gap different to the other a current will flow.

When a torque is applied to the shaft the air gap of the shaft transformer will change, resulting in a current flow. The indicator unit transformer air gap is then adjusted until no current flows. The air gaps in both transformers must now be exactly equal. The applied torque is directly proportional to the width of the air gap or the micrometer screw movement. Shaft power is found by multiplying the micrometer screw reading by the shaft speed and a constant for the meter.

Viscosity measurement

Viscosity control of fuels is essential if correct atomisation and combustion is to take place. Increasing the temperature of a fuel will

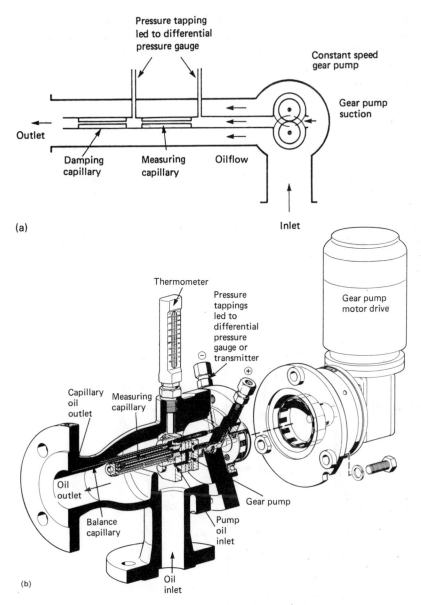

Figure 15.20 Viscosity sensor, (a) diagrammatic; (b) actual

reduce its viscosity, and vice-versa. As a result of the varying properties of marine fuels, often within one tank, actual viscosity must be continuously measured and then corrected by temperature adjustment. The sensing device is shown in Figure 15.20. A small constant speed gear pump forces a fixed quantity of oil through a capillary (narrow bore) tube. The liquid flow in the capillary is such that the difference in pressure readings taken before the capillary and after it is related to the oil viscosity. A differential pressure gauge is calibrated to read viscosity and the pressure values are used to operate the heater control to maintain some set viscosity value.

Salinometer

Water purity, in terms of the absence of salts, is essential where it is to be used as boiler feed. Pure water has a high resistance to the flow of electricity whereas salt water has a high electrical conductivity. A measure of conductivity, in siemens, is a measure of purity.

The salinity measuring unit shown in Figure 15.21 uses two small cells each containing a platinum and a gunmetal electrode. The liquid sample passes through the two cells and any current flow as a result of conductance is measured. Since conductivity rises with temperature a

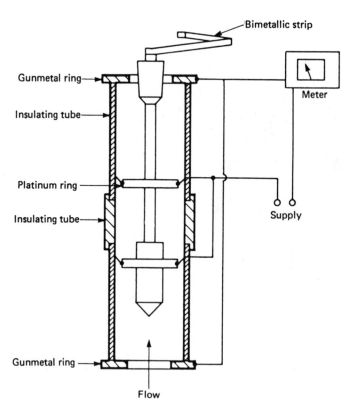

Figure 15.21 Salinometer

compensating resistor is incorporated in the measuring circuit. The insulating plunger varies the water flow in order to correct values to 20°C for a convenient measuring unit, the microsiemens/cm^3 or dionic unit. A de-gassifier should be fitted upstream of this unit to remove dissolved carbon dioxide which will cause errors in measurement.

Oxygen analyser

The measuring of oxygen content in an atmosphere is important, particularly when entering enclosed spaces. Also inert gas systems use exhaust gases which must be monitored to ensure that their oxygen content is below 5%. One type of instrument used to measure oxygen content utilises the fact that oxygen is attracted by a magnetic field, that is, it is paramagnetic.

A measuring cell uses a dumb-bell shaped wire which rotates in a magnetic field. The presence of oxygen will affect the magnetic field and cause rotation of the dumb-bell. The current required to align the dumb-bell is a measure of the oxygen concentration in the cell.

The sampling system for an inert gas main is shown in Figure 15.22. The probe at the tap-off point has an integral filter to remove dust. The

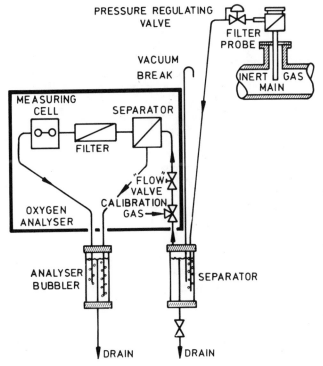

Figure 15.22 Oxygen analyser

gas then passes through a separator, a three-way valve and a flow valve. The gas sample, after further separation and filtering, passes to the measuring cell and part of it is bypassed. The flow valve is used to obtain the correct flow through the measuring cell and a meter provides the reading of oxygen content. The three-way valve permits the introduction of a zeroing gas (nitrogen) and a span gas (air). The span gas gives a 21% reading as a calibration check.

Oil-in-water monitor

Current regulations with respect to the discharge of oily water set limits of concentration between 15 and 100 parts per million. A monitor is required in order to measure these values and provide both continuous records and an alarm where the permitted level is exceeded.

The principle used is that of ultra-violet fluorescence. This is the emission of light by a molecule that has absorbed light. During the short interval between absorption and emission, energy is lost and light of a longer wavelength is emitted. Oil fluoresces more readily than water and this provides the means for its detection.

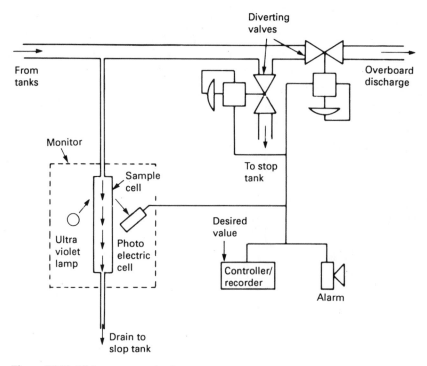

Figure 15.23 Oil-in-water monitoring system

A sample is drawn off from the overboard discharge and passes through a sample cell (Figure 15.23). An ultra-violet light is directed at the sample and the fluorescence is monitored by a photoelectric cell. The measured value is compared with the maximum desired value in the controller/recorder. Where an excessive level of contamination is detected an alarm is sounded and diverting valves are operated. The discharging liquid is then passed to a slop tank.

Control theory

To control a device or system is to be able to adjust or vary the parameters which affect it. This can be achieved manually or automatically, depending upon the arrangements made in the system. All forms of control can be considered to act in a loop. The basic elements present in the loop are a detector, a comparator/controller and a correcting unit, all of which surround the process and form the loop (Figure 15.24). This arrangement is an automatic closed loop if the elements are directly connected to one another and the control action takes place without human involvement. A manual closed loop would exist if one element were replaced by a human operator.

It can be seen therefore that in a closed loop control system the control action is dependent on the output. A detecting or measuring element will obtain a signal related to this output which is fed to the transmitter. From the transmitter the signal is then passed to a comparator. The comparator will contain some set or desired value of the controlled

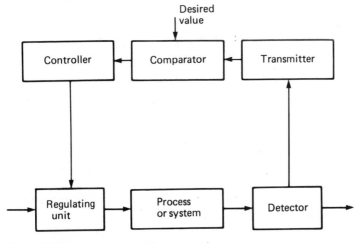

Figure 15.24 Automatic closed loop control

condition which is compared to the measured value signal. Any deviation or difference between the two values will result in an output signal to the controller. The controller will then take action in a manner related to the deviation and provide a signal to a correcting unit. The correcting unit will then increase or decrease its effect on the system to achieve the desired value of the system variable. The comparator is usually built in to the controller unit.

The transmitter, controller and regulating unit are supplied with an operating medium in order to function. The operating medium may be compressed air, hydraulic oil or electricity. For each medium various types of transmitting devices, controllers and regulating units are used.

Transmitters

Pneumatic

Many pneumatic devices use a nozzle and flapper system to give a variation in the compressed air signal. A pneumatic transmitter is shown in Figure 15.25. If the flapper moves *away* from the nozzle then the transmitted or output pressure will fall to a low value. If the flapper moves *towards* the nozzle then the transmitted pressure will rise to almost the supply pressure. The transmitted pressure is approximately proportional to the movement of the flapper and thus the change in the measured variable. The flapper movement will be very minute and where measurement of a reasonable movement is necessary a system of

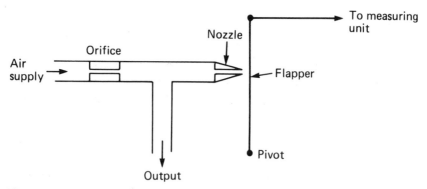

Figure 15.25 Position balance transmitter

levers and linkages must be introduced. This in turn leads to errors in the system and little more than on-off control.

Improved accuracy is obtained when a feedback bellows is added to assist in flapper positioning (Figure 15.26). The measured value acts on one end of the pivoted flapper against an adjustable spring which enables the measuring range to be changed. The opposite end of the flapper is acted upon by the feedback bellows and the nozzle. In operation a change in the measured variable may cause the flapper to approach the nozzle and thus build up the output signal pressure. The pressure in the feedback bellows also builds up, tending to push the flapper away from the nozzle, i.e. a negative feedback. An equilibrium

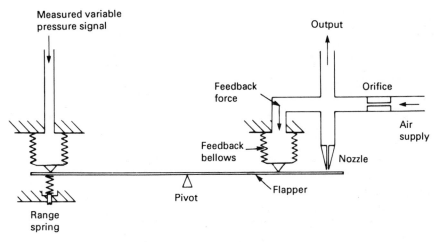

Figure 15.26 Force balance transmitter with feedback

position will be set up giving an output signal corresponding to the measured variable.

Most pneumatic transmitters will have relays fitted which magnify or amplify the output signals to reduce time lags in the system and permit signal transmission over considerable distances. Relays can also be used for mathematical operations, such as adding, subtracting, multiplying or dividing of signals. Such devices are known as 'summing' or 'computing relays'.

Electrical

Simple electrical circuits may be used where the measured variable causes a change in resistance which is read as a voltage or current and displayed in its appropriate units.

Another method is where the measured variable in changing creates a potential difference which, after amplification, drives a reversible motor to provide a display and in moving also reduces the potential difference to zero.

Alternating current positioning motors can be used as transmitters when arranged as shown in Figure 15.27. Both rotors are supplied from the same supply source. The stators are star wound and when the two rotor positions coincide there is no current flow since the e.m.f.s of both are equal and opposite. When the measured variable causes a change in

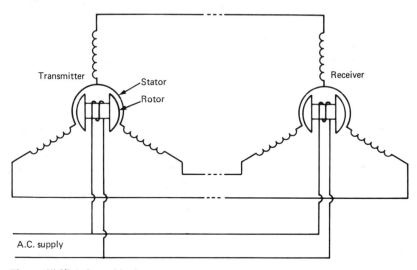

Figure 15.27 A.C. positioning motors

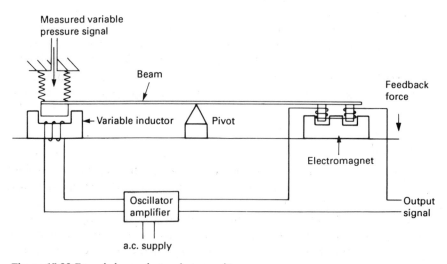

Figure 15.28 Force balance electronic transmitter

the transmitter rotor position, the two e.m.f.s will be out of balance. A current will flow and the receiver rotor will turn until it aligns with the transmitter. The receiver rotor movement will provide a display of the measured variable.

An electrical device can also be used as a transmitter (Figure 15.28). The measured variable acts on one end of a pivoted beam causing a change in a magnetic circuit. The change in the magnetic circuit results in a change in output current from the oscillator amplifier, and the oscillator output current operates an electromagnet so that it produces a negative feedback force which opposes the measured variable change. An equilibrium position results and provides an output signal.

Hydraulic

The telemotor of a hydraulically actuated steering gear is one example of a hydraulic transmitter. A complete description of the unit and its operation is given in Chapter 12.

Controller action

The transmitted output signal is received by the controller which must then undertake some corrective action. There will however be various time lags or delays occurring during first the measuring and then the transmission of a signal indicating a change. A delay will also occur in the action of the controller. These delays produce what is known as the transfer function of the unit or item, that is, the relationship between the output and input signals.

The control system is designed to maintain some output value at a constant desired value, and a knowledge of the various lags or delays in the system is necessary in order to achieve the desired control. The controller must therefore rapidly compensate for these system variations and ensure a steady output as near to the desired value as practicable.

Two-step or on-off

In this, the simplest of controller actions, two extreme positions of the controller are possible, either on or off. If the controller were, for example, a valve it would be either open or closed. A heating system is considered with the control valve regulating the supply of heating steam. The controller action and system response is shown in Figure 15.29. As the measured value rises above its desired value the valve will close. System lags will result in a continuing temperature rise which eventually peaks and then falls below the desired value. The valve will then open

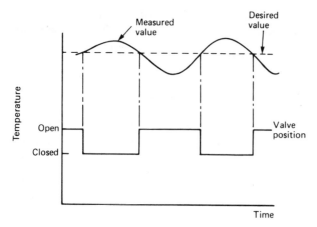

Figure 15.29 Two-step or on-off control

again and the temperature will cease to fall and will rise again. This form of control is acceptable where a considerable deviation from the desired value is allowed.

Continuous action

Proportional action

This is a form of continuous control where any change in controller output is proportional to the deviation between the controlled condition and the desired value. The *proportional band* is the amount by which the input signal value must change to move the correcting unit between its extreme positions. The desired value is usually located at the centre of the proportional band. *Offset* is a sustained deviation as a result of a load change in the process. It is an inherent characteristic of proportional control action. Consider, for example, a proportional controller operating a feedwater valve supplying a boiler drum. If the steam demand, i.e. load, increases then the drum level will fall. When the level has dropped the feedwater valve will open. An equilibrium position will be reached when the feedwater valve has opened enough to match the new steam demand. The drum level, however, will have fallen to a new value below the desired value, i.e. offset. See Figure 15.30.

Integral action

This type of controller action is used in conjunction with proportional control in order to remove offset. Integral or reset action occurs when the controller output varies at a rate proportional to the deviation

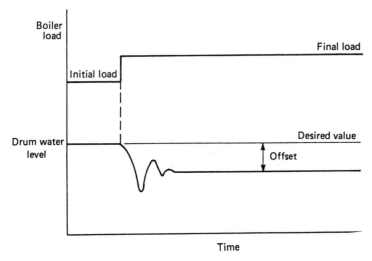

Figure 15.30 System response to proportional controller action

between the desired value and the measured value. The integral action of a controller can usually be varied to achieve the required response in a particular system.

Derivative action

Where a plant or system has long time delays between changes in the measured value and their correction, derivative action may be applied. This will be in addition to proportional and integral action. Derivative or rate action is where the output signal change is proportional to the rate of change of deviation. A considerable corrective action can therefore take place for a small deviation which occurs suddenly. Derivative action can also be adjusted within the controller.

Multiple-term controllers

The various controller actions in response to a process change are shown in Figure 15.31. The improvement in response associated with the addition of integral and derivative action can clearly be seen. Reference is often made to the number of terms of a controller. This means the various actions: proportional (P), integral (I), and derivative (D). A three-term controller would therefore mean P+I+D, and two-term usually P+I. A controller may be arranged to provide either split range or cascade control, depending upon the arrangements in the control system. These two types of control are described in the section dealing with control systems.

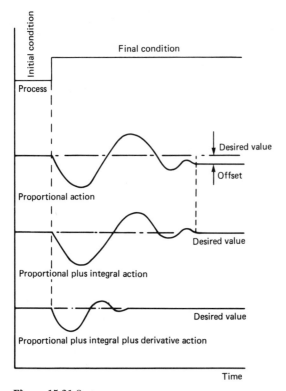

Figure 15.31 System response

Controllers

The controller may be located close to the variable measuring point and thus operate without the use of a transmitter. It may however be located in a remote control room and receive a signal from a transmitter and relay, as mentioned earlier. The controller is required to maintain some system variable at a desired value regardless of load changes. It may also indicate the system variable and enable the desired value to be changed. Over a short range about the desired value the controller will generate a signal to operate the actuating mechanism of the correcting unit. This control signal may include proportional, integral and derivative actions, as already described. Where all three are used it may be known as a 'three-term' controller.

A pneumatic three-term controller is shown in Figure 15.32. Any variation between desired and measured values will result in a movement of the flapper and change in the output pressure. If the

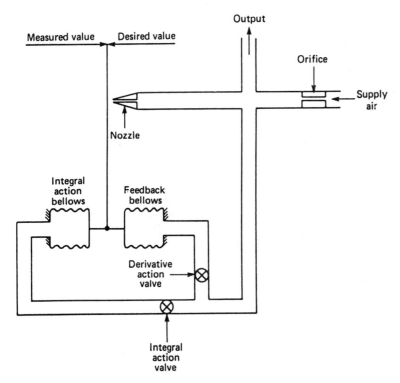

Figure 15.32 Pneumatic three-term controller

derivative action valve is open and the integral action valve closed, then only proportional control occurs. It can be seen that as the flapper moves towards the nozzle a pressure build-up will occur which will increase the output pressure signal and also move the bellows so that the flapper is moved away from the nozzle. This is then a negative feedback which is proportional to the flapper or measured value movement. When the integral action valve is opened, any change in output signal pressure will affect the integral action bellows which will oppose the feedback bellows movement. Varying the opening of the integral action valve will alter the amount of integral action of the controller. Closing the derivative action valve any amount would introduce derivative action. This is because of the delay that would be introduced in the provision of negative feedback for a sudden variable change which would enable the output signal pressure to build up. If the measured variable were to change slowly then the proportional action would have time to build up and thus exert its effect.

An electronic three-term controller is shown in Figure 15.33. The controller output signal is subjected to the various control actions, in this

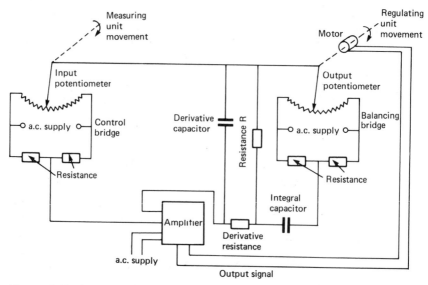

Figure 15.33 Electronic three-term controller

case by electronic components and suitable circuits. Any change in the measured value will move the input potentiometer and upset the balance of the control bridge. A voltage will then be applied to the amplifier and the amplifier will provide an output signal which will result in a movement of the output potentiometer. The balancing bridge will then provide a voltage to the amplifier which equals that from the control bridge. The amplifier output signal will then cease.

The output potentiometer movement is proportional to the deviation between potentiometer positions, and movement will continue while the deviation remains. Integral and derivative actions are obtained by the resistances and capacitors in the circuit. With the integral capacitor fitted while a deviation exists there will be a current through resistance R as a result of the voltage across it. This current flow will charge the integral capacitor and thus reduce the voltage across R. The output potentiometer must therefore continue moving until no deviation exists. No offset can occur as it would if only proportional action took place. Derivative action occurs as a result of current flow through the derivative resistor which also charges the derivative capacitor. This current flow occurs only while the balancing bridge voltage is changing, but a larger voltage is required because of the derivative capacitor. The derivative action thus results in a faster return to the equilibrium position, as would be expected. The output potentiometer is moved by a motor which also provides movement for a valve or other correcting unit in the controlled process.

Correcting unit

The controller output signal is fed to the correcting unit which then alters some variable in order to return the system to its desired value. This correcting unit may be a valve, a motor, a damper or louvre for a fan or an electric contactor. Most marine control applications will involve the actuation or operation of valves in order to regulate liquid flow.

Pneumatic control valve

A typical pneumatic control valve is shown in Figure 15.34. It can be considered as made up of two parts—the actuator and the valve. In the arrangement shown a flexible diaphragm forms a pressure tight chamber in the upper half of the actuator and the controller signal is fed in. Movement of the diaphragm results in a movement of the valve spindle and the valve. The diaphragm movement is opposed by a spring and is usually arranged so that the variation of controller output corresponds to full travel of the valve.

The valve body is arranged to fit into the particular pipeline and houses the valve and seat assembly. Valve operation may be direct acting where increasing pressure on the diaphragm closes the valve. A reverse acting valve opens as pressure on the diaphragm increases. The diaphragm movement is opposed by a spring which will close or open the valve in the event of air supply failure depending upon the action of the valve.

The valve disc or plug may be single or double seated and have any of a variety of shapes. The various shapes and types are chosen according to the type of control required and the relationship between valve lift and liquid flow.

A non-adjustable gland arrangement is usual. Inverted V-ring packing is used to minimise the friction against the moving spindle.

In order to achieve accurate valve disc positioning and overcome the effects of friction and unbalanced forces a valve positioner may be used. The operating principle is shown in Figure 15.35. The controller signal acts on a bellows which will move the flapper in relation to the nozzle. This movement will alter the air pressure on the diaphragm which is supplied via an orifice from a constant pressure supply. The diaphragm movement will move the valve spindle and also the flapper. An equilibrium position will be set up when the valve disc is correctly positioned. This arrangement enables the use of a separate power source to actuate the valve.

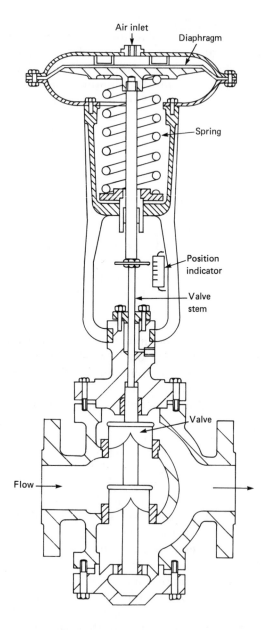

Figure 15.34 Pneumatically controlled valve

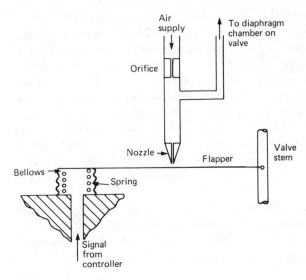

Figure 15.35 Valve positioner

Actuator operation

The control signal to a correcting unit may be pneumatic, electric or hydraulic. The actuating power may also be any one of these three and not necessarily the same as the control medium.

Electrical control signals are usually of small voltage or current values which are unable to effect actuator movement. Pneumatic or hydraulic power would then be used for actuator operation.

A separate pneumatic power supply may be used even when the control signal is pneumatic, as described in the previous section.

Hydraulic actuator power is used where large or out of balance forces occur or when the correcting unit is of large dimensions itself. Hydraulic control with separate hydraulic actuation is a feature of some types of steering gear, as mentioned in Chapter 12.

Control systems

Boiler water level

A modern high-pressure, high-temperature watertube boiler holds a small quantity of water and produces large quantities of steam. Very careful control of the drum water level is therefore necessary. The reactions of steam and water in the drum are complicated and require a control system based on a number of measured elements.

When a boiler is operating the water level in the gauge glass reads higher than when the boiler is shut down. This is because of the presence of steam bubbles in the water, a situation which is accepted in normal practice. If however there occurs a sudden increase in steam demand from the boiler the pressure in the drum will fall. Some of the water present in the drum at the higher pressure will now 'flash off' and become steam. These bubbles of steam will cause the drum level to rise. The reduced mass of water in the drum will also result in more steam being produced, which will further raise the water level. This effect is known as 'swell'. A level control system which used only level as a measuring element would close in the feed control valve—when it should be opening it.

When the boiler load returns to normal the drum pressure will rise and steam bubble formation will reduce, causing a fall in water level. Incoming colder feed water will further reduce steam bubble formation and what is known as 'shrinkage' of the drum level will occur.

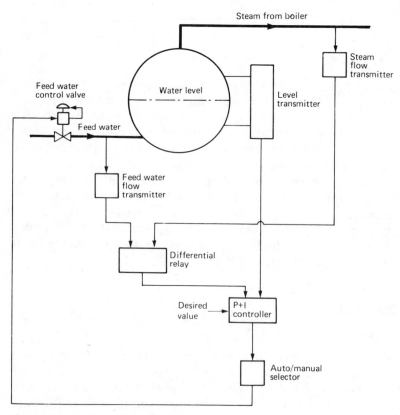

Figure 15.36 Boiler water level control

The problems associated with swell and shrinkage are removed by the use of a second measuring element, 'steam flow'. A third element, 'feed water flow', is added to avoid problems that would occur if the feed water pressure were to vary.

A three element control system is shown in Figure 15.36. The measured variables or elements are 'steam flow', 'drum level' and 'feed water flow'. Since in a balanced situation steam flow must equal feed flow, these two signals are compared in a differential relay. The relay output is fed to a two-term controller and comparator into which the measured drum level signal is also fed. Any deviation between the desired and actual drum level and any deviation between feed and steam flow will result in controller action to adjust the feed water control valve. The drum level will then be returned to its correct position.

A sudden increase in steam demand would result in a deviation signal from the differential relay and an output signal to open the feed water control valve. The swell effect would therefore not influence the correct operation of the control system. For a reduction in steam demand, an output signal to close the feedwater control valve would result, thus avoiding shrinkage effects. Any change in feed water pressure would result in feed water control valve movement to correct the change before the drum level was affected.

Exhaust steam pressure control

Exhaust steam for various auxiliary services may be controlled at constant pressure by appropriate operation of a surplus steam (dump) valve or a make-up steam valve. A single controller can be used to operate one valve or the other in what is known as 'split range control'.

The control arrangement is shown in Figure 15.37. The steam pressure in the auxiliary range is measured by a pressure transmitter.

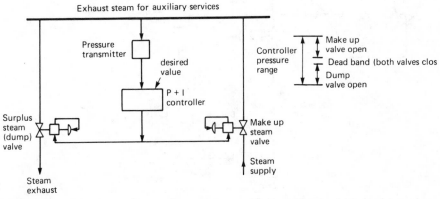

Figure 15.37 Exhaust steam pressure control

This signal is fed to the controller where it is compared with the desired value. The two-term controller will provide an output signal which is fed to both control valves. Each valve is operated by a different range of pressure with a 'dead band' between the ranges so that only one valve is ever open at a time. The arrangement is shown in Figure 15.37. Thus if the auxiliary range pressure is high the dump valve opens to release steam. If the pressure is low the make-up valve opens to admit steam.

This split range control principle can be applied to a number of valves if the controller output range is split appropriately.

Steam temperature control

Steam temperature control of high pressure superheated steam is necessary to avoid damage to the metals used in a steam turbine.

One method of control is shown in Figure 15.38. Steam from the primary superheater may be directed to a boiler drum attemperator where its temperature will be reduced. This steam will then be further heated in the secondary superheater. The steam temperature leaving the secondary superheater is measured and transmitted to a three-term

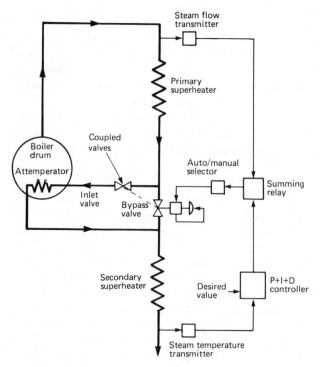

Figure 15.38 Steam temperature control

controller which also acts as a comparator. Any deviation from the desired value will result in a signal to a summing relay. The other signal to the relay is from a steam flow measuring element. The relay output signal provides control of the coupled attemperator inlet and bypass valves. As a result the steam flow is proportioned between the attemperator and the straight through line. This two-element control system can adequately deal with changing conditions. If, for example, the steam demand suddenly increased a fall in steam temperature might occur. The steam flow element will however detect the load change and adjust the amount of steam attemperated to maintain the correct steam temperature.

Boiler combustion control

The essential requirement for a combustion control system is to correctly proportion the quantities of air and fuel being burnt. This will ensure complete combustion, a minimum of excess air and acceptable exhaust gases. The control system must therefore measure the flow rates of fuel oil and air in order to correctly regulate their proportions.

A combustion control system capable of accepting rapid load changes is shown in Figure 15.39. Two control elements are used, 'steam flow' and 'steam pressure'. The steam pressure signal is fed to a two-term controller and is compared with the desired value. Any deviation results in a signal to the summing relay.

The steam flow signal is also fed into the summing relay. The summing relay which may add or subtract the input signals provides an output which represents the fuel input requirements of the boiler. This output becomes a variable desired value signal to the two-term controllers in the fuel control and combustion air control loops. A high or low signal selector is present to ensure that when a load change occurs the combustion air flow is always in excess of the fuel requirements. This prevents poor combustion and black smokey exhaust gases. If the master signal is for an increase in steam flow, then when it is fed to the low signal selector it is blocked since it is the higher input value. When the master signal is input to the high signal selector it passes through as the higher input. This master signal now acts as a variable desired value for the combustion air sub-loop and brings about an increased air flow. When the increased air flow is established its measured value is now the higher input to the low signal selector. The master signal will now pass through to bring about the increased fuel supply to the boiler via the fuel supply sub-loop. The air supply for an increase in load is therefore established before the increase in fuel supply occurs. The required air to fuel ratio is set in the ratio relay in the air flow signal lines.

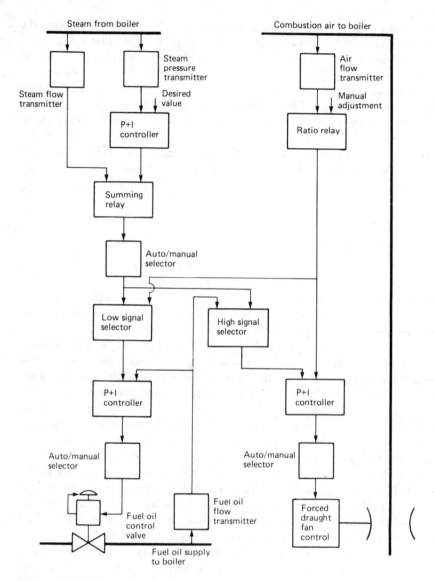

Figure 15.39 Boiler combustion control

Cooling water temperature control

Accurate control of diesel engine cooling water temperature is a requirement for efficient operation. This can be achieved by a single controller under steady load conditions, but because of the fluctuating situation during manœuvring a more complex system is required.

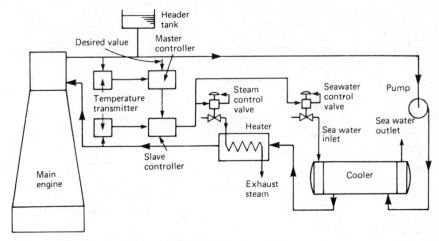

Figure 15.40 Cooling water temperature control

The control system shown in Figure 15.40 uses a combination of cascade and split range control. Cascade control is where the output from a master controller is used to automatically adjust the desired value of a slave controller. The master controller obtains an outlet temperature reading from the engine which is compared with a desired value. Any deviation acts to adjust the desired value of the slave controller. The slave controller also receives a signal from the water inlet temperature sensor which it compares with its latest desired value. Any deviation results in a signal to two control valves arranged for split range control. If the cooling water temperature is high, the sea water valve is opened to admit more cooling water to the cooler. If the cooling water temperature is low, then the sea water valve will be closed in. If the sea water valve is fully closed, then the steam inlet valve to the water heater will be opened to heat the water. Both master and slave controllers will be identical instruments and will be two-term (P+I) in action.

Another method of temperature control involves the use of only a single measuring element (Figure 15.41). A three-way valve is provided in the cooling water line to enable bypassing of the cooler. The cooler is provided with a full flow of sea water which is not controlled by the system. A temperature sensing element on the water outlet provides a signal to a two-term controller (P+I). The controller is provided with a desired value and any deviation between it and the signal will result in an output to the three-way control valve. If the measured temperature is low, more water will be bypassed and its temperature will therefore increase. If the measured temperature is high, then less water will be bypassed, more will be cooled and the temperature will fall. A simple

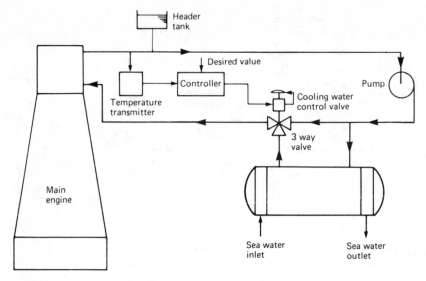

Figure 15.41 Cooling water temperature control

system such as this can be used only after careful analysis of the plant conditions and the correct sizing of equipment fitted.

Centralised control

The automatic control concept, correctly developed, results in the centralising of control and supervisory functions. All ships have some degree of automation and instrumentation which is centred around a console. Modern installations have machinery control rooms where the monitoring of control functions takes place. The use of a separate room in the machinery space enables careful climate control of the space for the dual benefit of the instruments and the engineer.

Control consoles are usually arranged with the more important controls and instrumentation located centrally and within easy reach. The display panels often make use of mimic diagrams. These are line diagrams of pipe systems or items of equipment which include miniature alarm lights or operating buttons for the relevant point or item in the system. A high-temperature alarm at, for instance, a particular cylinder exhaust would display at the appropriate place on the mimic diagram of the engine. Valves shown on mimic diagrams would be provided with an indication of their open or closed position, pumps would have a running light lit if operating, etc. The grouping of the controllers and instrumentation for the various systems previously described enables them to become part of the complete control system for the ship.

The ultimate goal in the centralised control room concept will be to perform and monitor every possible operation remotely from this location. This will inevitably result in a vast amount of information reaching the control room, more than the engineer supervisor might reasonably be expected to continuously observe. It is therefore usual to incorporate data recording and alarm systems in control rooms. The alarm system enables the monitoring of certain measured variables over a set period and the readings obtained are compared with some reference or desired value. Where a fault condition is located, i.e. a measured value different from the desired value, audible and visual alarms are given and a print-out of the fault and the time of occurrence is produced. Data recording or data logging is the production of measured variable information either automatically at set intervals or on demand. A diagrammatic layout of a data logging and alarm monitoring system is shown in Figure 15.42.

Unattended machinery spaces

The sophistication of modern control systems and the reliability of the equipment used have resulted in machinery spaces remaining un-attended for long periods. In order to ensure the safety of the ship and its equipment during UMS operation certain essential requirements must be met:

1. *Bridge control.* A control system to operate the main machinery must be provided on the bridge. Instrumentation providing certain basic information must be provided.
2. *Machinery control room.* A centralised control room must be provided with the equipment to operate all main and auxiliary machinery easily accessible.
3. *Alarm and fire protection.* An alarm system is required which must be comprehensive in coverage of the equipment and able to provide warnings in the control room, the machinery space, the accommodation and on the bridge. A fire detection and alarm system which operates rapidly must also be provided throughout the machinery space, and a fire control point must be provided outside the machinery space with facilities for control of emergency equipment.
4. *Emergency power.* Automatic provision of electrical power to meet the varying load requirements. A means of providing emergency electrical power and essential lighting must be provided. This is usually met by the automatic start up of a standby generator.

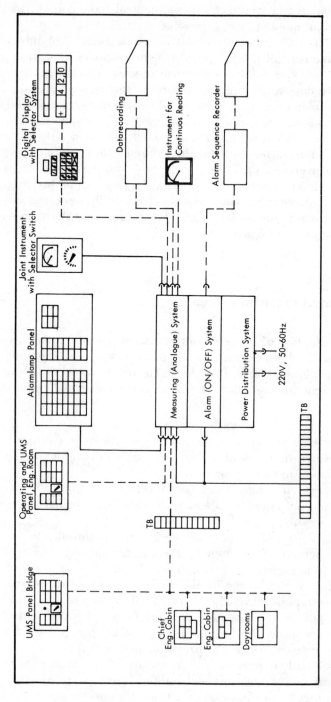

Figure 15.42 Data logging and alarm monitoring system

Bridge control

Equipment operation from the machinery control room will be by a trained engineer. The various preparatory steps and logical timed sequence of events which an engineer will undertake cannot be expected to occur when equipment is operated from the bridge. Bridge control must therefore have built into the system appropriate circuits to provide the correct timing, logic and sequence. There must also be protection devices and safety interlocks built into the system.

A bridge control system for a steam turbine main propulsion engine is shown in Figure 15.43. Control of the main engine may be from the bridge control unit or the machinery control room. The programming and timing unit ensures that the correct logical sequence of events occurs over the appropriate period. Typical operations would include the raising of steam in the boiler, the circulating of lubricating oil through the turbine and the opening of steam drains from the turbine.

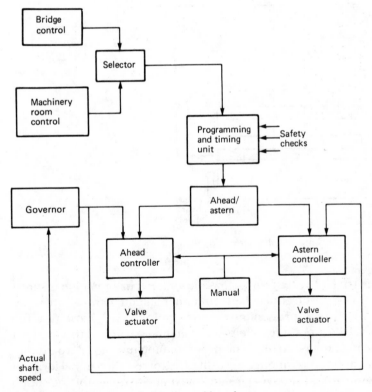

Figure 15.43 Bridge control of steam turbine plant

The timing of certain events, such as the opening and closing of steam valves, must be carefully controlled to avoid dangerous conditions occurring or to allow other system adjustments to occur. Protection and safety circuits or interlocks would be input to the programming and timing unit to stop its action if, for example, the turning gear was still engaged or the lubricating oil pressure was low. The ahead/astern selector would direct signals to the appropriate valve controller resulting in valve actuation and steam supply. When manœuvring some switching arrangement would ensure that the astern guardian valve was open, bled steam was shut off, etc. If the turbine were stopped it would automatically receive blasts of steam at timed intervals to prevent rotor distortion. A feedback signal of shaft speed would ensure correct speed without action from the main control station.

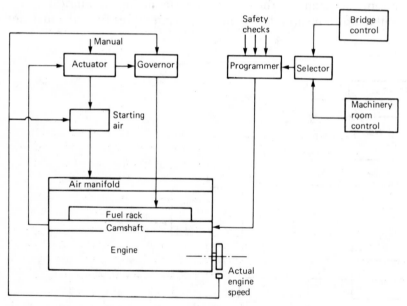

Figure 15.44 Bridge control of slow-speed diesel engine

A bridge control system for a slow-speed diesel main engine is shown in Figure 15.44. Control may be from either station with the operating signal passing to a programming and timing unit. Various safety interlocks will be input signals to prevent engine starting or to shut down the engine if a fault occurred. The programming unit signal would then pass to the camshaft positioner to ensure the correct directional location. A logic device would receive the signal next and arrange for the supply of starting air to turn the engine. A signal passing through the governor

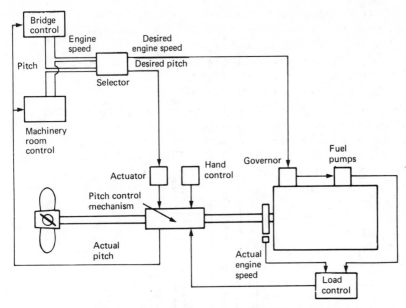

Figure 15.45 Bridge control of controllable-pitch propeller

would supply fuel to the engine to start and continue operation. A feedback signal of engine speed would shut off the starting air and also enable the governor to control engine speed. Engine speed would also be provided as an instrument reading at both control stations.

A bridge control system for a controllable-pitch propeller is shown in Figure 15.45. The propeller pitch and engine speed are usually controlled by a single lever (combinator). The control lever signal passes via the selector to the engine governor and the pitch-operating actuator. Pitch and engine speed signals will be fed back and displayed at both control stations. The load control unit ensures a constant load on the engine by varying propeller pitch as external conditions change. The input signals are from the fuel pump setting and actual engine speed. The output signal is supplied as a feedback to the pitch controller.

The steering gear is, of course, bridge controlled and is arranged for automatic or manual control. A typical automatic or auto pilot system is shown in Figure 15.46. A three-term controller provides the output signal where a course deviation exists and will bring about a rudder movement. The various system parts are shown in terms of their system functions and the particular item of equipment involved. The feedback loop between the rudder and the amplifier (variable delivery pump) results in no pumping action when equilibrium exists in the system. External forces can act on the ship or the rudder to cause a change in the

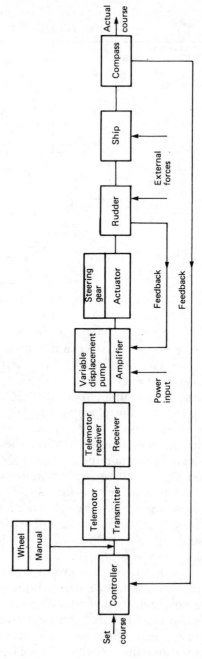

Figure 15.46 Automatic steering system

ship's actual course resulting in a feedback to the controller and subsequent corrective action. The controller action must be correctly adjusted for the particular external conditions to ensure that excessive rudder movement does not occur.

Electrical supply control

The automatic provision of electrical power to meet varying load demands can be achieved by performing the following functions automatically:

1. Prime mover start up.
2. Synchronising of incoming machine with bus-bars.
3. Load sharing between alternators.
4. Safety and operational checks on power supply and equipment in operation.
5. Unloading, stopping and returning to standby of surplus machines.
6. Preferential tripping of non-essential loads under emergency conditions and their reinstating when acceptable.

A logic flow diagram for such a system is given in Figure 15.47. Each of three machines is considered able to supply 250 kW. A loading in excess of this will result in the start up and synchronising of another

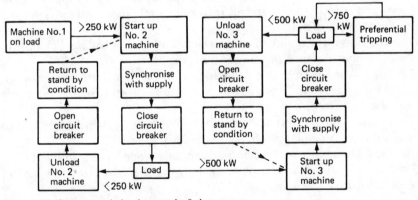

Figure 15.47 Automatic load control of alternators

machine. Should the load fall to a value where a running machine is unnecessary it will be unloaded, stopped and returned to the standby condition. If the system should overload through some fault, such as a machine not starting, an alarm will be given and preferential tripping will occur of non-essential loads. Should the system totally fail the emergency alternator will start up and supply essential services and lighting through its switchboard.

Chapter 16
Engineering materials

A knowledge of the properties of a material is essential to every engineer. This enables suitable material choice for a particular application, appropriate design of the components or parts, and their protection, where necessary, from corrosion or damage.

Material properties

The behaviour of a metal under various conditions of loading is often described by the use of certain terms:

Tensile strength. This is the main single criterion with reference to metals. It is a measure of the material's ability to withstand the loads upon it in service. Terms such as 'stress', 'strain', 'ultimate tensile strength', 'yield stress' and 'proof stress' are all different methods of quantifying the tensile strength of the material.

Ductility. This is the ability of a material to undergo permanent change in shape without rupture or loss of strength.

Brittleness. A material that is liable to fracture rather than deform when absorbing energy (such as impact) is said to be brittle. Strong materials may also be brittle.

Malleability. A material that can be shaped by beating or rolling is said to be malleable. A similar property to ductility.

Plasticity. The ability to deform permanently when load is applied.

Elasticity. The ability to return to the original shape or size after having been deformed or loaded.

Toughness. A combination of strength and the ability to absorb energy or deform plastically. A condition between brittleness and softness.

Hardness. A material's ability to resist plastic deformation usually by indentation.

Testing of materials

Various tests are performed on materials in order to quantify their properties and determine their suitability for various engineering applications. For measurement purposes a number of terms are used, with 'stress' and 'strain' being the most common. Stress, or more correctly 'intensity of stress', is the force acting on a unit area of the material. Strain is the deforming of a material due to stress. When a force is applied to a material which tends to shorten or compress it, the stress is termed 'compressive stress'. When the force applied tends to lengthen the material it is termed 'tensile stress'. When the force tends to cause the various parts of the material to slide over one another the stress is termed 'shear stress'.

Tensile test

A tensile test measures a material's strength and ductility. A specially shaped specimen of standard size is gripped in the jaws of a testing machine, and a load gradually applied to draw the ends of the specimen apart such that it is subject to tensile stress. The original test length of the specimen, L_1, is known and for each applied load the new length, L_2, can be measured. The specimen will be found to have extended by some small amount, $L_2 - L_1$. This deformation, expressed as

$$\frac{\text{extension}}{\text{original length}}$$

is known as the linear strain.

Additional loading of the specimen will produce results which show a uniform increase of extension until the yield point is reached. Up to the yield point or elastic limit, the removal of load would have resulted in the specimen returning to its original size. The stress and strain values for various loads can be shown on a graph as in Figure 16.1. If testing

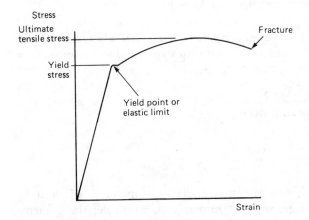

Figure 16.1 Stress strain curve

continues beyond the yield point the specimen will 'neck' or reduce in cross section. The load values divided by the original cross section would give the shape shown. The highest stress value is known as the 'ultimate tensile stress' (UTS) of the material.

Within the elastic limit, stress is proportional to strain, and therefore

$$\frac{\text{stress}}{\text{strain}} = \text{constant}$$

This constant is known as the 'modulus of elasticity' (E) of the material. The yield stress is the value of stress at the yield point. Where a clearly defined yield point is not obtained, a proof stress value is given. This is obtained by drawing a line parallel to the stress–strain line at a value of strain, usually 0.1%. The intersection of the two lines is considered the proof stress (Figure 16.2).

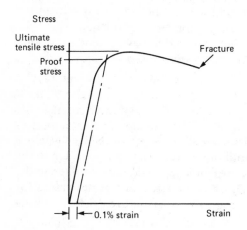

Figure 16.2 Stress strain curve—material without a definite yield point

A 'factor of safety' is often specified for materials where this is the ratio of ultimate tensile strength to working stress, and is always a value greater than unity.

$$\text{factor of safety} = \frac{\text{UTS}}{\text{working stress}}$$

Impact test

This test measures the energy absorbed by a material when it is fractured. There are a number of impact tests available; the Charpy vee-notch test is usually specified. The test specimen is a square section

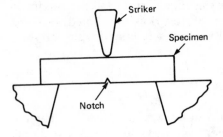

Figure 16.3 Impact test

bar with a vee-notch cut in the centre of one face. The specimen is mounted horizontally with the notch axis vertical (Figure 16.3). The test involves the specimen being struck opposite the notch and fractured. A striker or hammer on the end of a swinging pendulum provides the blow which breaks the specimen. The energy absorbed by the material in fracturing is measured by the machine.

Hardness test

The hardness test measures a material's resistance to indentation. A hardened steel ball or a diamond point is pressed onto the material surface for a given time with a given load. The hardness number is a function of the load and the area of the indentation. The value may be given as a Brinell number or Vickers Pyramid number, depending upon the machine used.

Creep test

Creep is the slow plastic deformation of a material under a constant stress. The test uses a specimen similar to that for a tensile test. A constant load is applied and the temperature is maintained constant. Accurate measurements of the increase in length are taken often over very long periods. The test is repeated for various loads and the material tested at what will be its temperature in service. Creep rate and limiting stress values can thus be found.

Fatigue test

Fatigue failure results from a repeatedly applied fluctuating stress which may be a lower value than the tensile strength of the material. A specially shaped specimen is gripped at one end and rotated by a fast revolving electric motor. The free end has a load suspended from it and a ball race is fitted to prevent the load from turning. The specimen, as it turns, is therefore subjected to an alternating tensile and compressive stress. The stress reversals are counted and the machine is run until the specimen breaks. The load and the number of reversals are noted and the procedure repeated. The results will provide a limiting fatigue stress or fatigue limit for the material.

Bend test

The bend test determines the ductility of a material. A piece of material is bent through 180° around a former. No cracks should appear on the material surface.

Non-destructive testing

A number of tests are available that do not damage the material under test and can therefore be used on the finished item if required. These tests are mainly examinations of the material to ensure that it is defect free and they do not, as such, measure properties.

Various penetrant liquids can be used to detect surface cracks. The penetrant liquid will be chosen for its ability to enter the smallest of cracks and remain there. A means of detecting the penetrant is then required which may be an ultra-violet light where a fluorescent penetrant is used. Alternatively a red dye penetrant may be used and after the surface is wiped clean, a white developer is applied.

Radiography, the use of X-rays or γ-rays to darken a photographic plate, can be used to detect internal flaws in materials. The shadow image produced will show any variations in material density, gas or solid inclusions, etc.

Ultrasonic testing is the use of high-frequency sound waves which reflect from the far side of the material. The reflected waves can be displayed on a cathode ray oscilloscope. Any defects will also result in reflected waves. The defect can be detected in size and location within the material.

Iron and steel production

Iron and steel are the most widely used materials and a knowledge of their manufacture and properties is very useful.

Making iron is the first stage in the production of steel. Iron ores are first prepared by crushing, screening and roasting with limestone and coke. The ore is thus concentrated and prepared for the blast furnace. A mixture of ore, coke and limestone is used to fill the blast furnace. Within the furnace an intense heat is generated as a result of the coke burning. Blasts of air entering the furnace towards the base assist in this burning process. The iron ore is reduced to iron and falls to the base of the furnace, becoming molten as it falls. Various impurities, such as carbon, silicon, manganese and sulphur, are absorbed by the iron as it descends. A slag of various materials, combined with the limestone, forms on top of the iron. The slag is tapped or drawn off from the furnace as it collects. The molten iron may be tapped and run into moulds to make bars of pig iron. Alternatively it may be transferred while molten to a steel manufacturing process.

Various processes are used in the manufacture of steel, such as the open hearth process, the oxygen or basic oxygen process and the electric furnace process. The terms 'acid' or 'basic' are often used with reference to steels. These terms refer to the production process and the type of furnace lining, e.g. an alkaline or basic lining is used to make basic steel. The choice of furnace lining is decided by the raw materials used in the manufacture of the steel. In all the steel producing processes the hot molten steel is exposed to air or oxygen which oxidises the impurities to refine the pig iron into high-quality steel.

Steels produced in the above processes will all contain an excess of oxygen which will affect the material quality. Several finishing treatments are used in the final steel casting. Rimmed steel has little or no oxygen removing treatment, and the central core of the solidified ingot is therefore a mass of blow holes. Hot rolling of the ingot usually welds up most of these holes. Killed steel is produced by adding aluminium or silicon before the molten steel is poured. The oxygen forms oxides with this material and a superior quality steel compared with rimmed steel is produced. Vacuum degassed steels result from reducing the atmospheric pressure while the steel is molten. This reduces the oxygen content and a final deoxidation can be achieved with small additions of silicon or aluminium.

Cast iron is produced by remelting pig iron under controlled conditions in a miniature type of blast furnace known as a 'cupola'. Variations of alloying additions may also be made. Two main types of cast iron occur—'white' and 'grey'. The colour relates to the appearance of the fractured surface. White cast iron is hard and brittle; grey is softer, readily machinable and less brittle.

Heat treatment

Heat treatment consists of heating a metal alloy to a temperature below its melting point and then cooling it in a particular manner. The result is some desired change in the material properties. Since most heat treatment is applied to steel, the various terms and types of treatment will be described with reference to steel.

Normalising. The steel is heated to a temperature of 850–950°C, depending upon its carbon content, and is then allowed to cool in air. A hard strong steel with a refined grain structure is produced.

Annealing. Again the steel is heated to around 850–950°C, but it is cooled slowly, either in the furnace or an insulated space. A softer, more ductile, steel than that in the normalised condition, is produced.

Hardening. The steel is heated to 850–950°C and is then rapidly cooled by quenching in oil or water. The hardest possible condition for the particular steel is thus produced and the tensile strength is increased.

Tempering. This process follows the quenching of steel and involves reheating to some temperature up to about 680°C. The higher the tempering temperature the lower the tensile properties of the material. Once tempered the metal is rapidly cooled by quenching.

Controlled rolling. This is sometimes described as a thermomechanical treatment. In two-stage controlled rolling an initial rough rolling is first carried out at 950–1100°C. The first controlled rolling stage is carried out at 850–920°C. The second stage is completed at about 700–730°C. The process is designed to achieve fine grain size, improve mechanical properties and toughness, and enhance weldability.

Material forming

In the production of engineering equipment various different processes are used to produce the assortment of component parts. These forming or shaping processes can be grouped as follows:

1. Casting.
2. Forging.
3. Extruding.
4. Sintering.
5. Machining.

Casting is the use of molten metal poured into a mould of the desired shape. A wooden pattern, slightly larger in dimensions than the desired item, to allow for shrinkage, may be used to form a mould in sand. Entry and exit holes, the gate and riser, are provided for the metal in the sand mould. Alternatively a permanent metal mould or 'die' may be made in two parts and used to make large quantities of the item. This method is called 'die casting'. The molten metal may be poured into the dies or forced in under pressure.

Forging involves shaping the metal when it is hot but not molten. In the manufacturing process of forging a pair of die blocks have the hot metal forced into them. This is usually achieved by placing the metal on the lower half die and forcing the top half down by a hydraulic press.

Extrusion involves the shaping of metal, usually into a rod or tube cross section, by forcing a block of material through appropriately shaped dies. Most metals must be heated before extrusion in order to reduce the extruding pressure required.

Sintering is the production of shaped parts from metal powder. A suitable metal powder mixture is placed in a die, compressed and heated to a temperature about two thirds of the material melting point. This heating process results in the powder compacting into a metal in the required shape.

Machining of one type or another is usually carried out on all metal items. This may involve planing flat surfaces, drilling holes, grinding rough edges, etc. Various equipment, such as milling machines, drilling machines, grinders, lathes, etc., will be used. Many of these machines are automatic or semi-automatic in operation and can perform a number of different operations in sequence.

Common metals and alloys

Some of the more common metals met in engineering will now be briefly described. Most metals are alloyed in order to combine the better qualities of the constituents and sometimes to obtain properties that none of them alone possesses. The various properties, composition and uses of some common engineering materials are given in Table 16.1.

Steel

Steel is an alloy of carbon and iron. Various other metals are alloyed to steel in order to improve the properties, reduce the heat treatment necessary and provide uniformity in large masses of the material. Manganese is added in amounts up to about 1.8% in order to improve mechanical properties. Silicon is added in amounts varying from 0.5% to 3.5% in order to increase strength and hardness. Nickel, when added as 3 to 3.75% of the content, produces a finer grained material with increased strength and erosion resistance. Chromium, when added, tends to increase grain size and cause hardness but improves resistance to erosion and corrosion. Nickel and chromium added to steel as 8% and 18% respectively produce stainless steel. Molybdenum is added in small amounts to improve strength, particularly at high temperatures. Vanadium is added in small amounts to increase strength and resistance to fatigue. Tungsten added at between 12 and 18%, together with up to 5% chromium, produces high speed steel.

Table 16.1 Material properties and uses

Material	Composition	0.2% proof stress (MN/m²)	Ultimate tensile stress (MN/m²)	% Elongation	Modulus of elasticity (MN/m²)	Brinell hardness number	Fatigue limit (MN/m²)	Uses
Admiralty brass	70 Cu, 29 Zn, 1 Sn	170	408	35	103	100	127	Tube plates and tubes for condensers
Aluminium	Almost pure	20	55	55	73	15	31	Base metal for light alloys
Aluminium brass	76 Cu, 22 Zn, 2 Al	139	378	55	110	75	96	Tube plates and tubes for condensers and heat exchangers
Brass	70 Cu, 30 Zn	115	324	67	115	65	106	Bearing liners
Cast iron (grey)	3.25 C, 2.25 Si, 0.65 Mn, Remainder Fe	180	310	5	120	200	110	Cylinder heads and liners
Copper	Almost pure	48	216	48	117	42	117	Base metal for numerous alloys
Cupro-nickel	70 Cu, 30 Ni	170	417	42	152	90	147	Heat exchanger tubes
Gunmetal	88 Cu, 10 Sn, 2 Zn	139	286	18	97	85	100	Valves and bearing bushes
Monel metal	68 Ni, 29 Cu, Remainder Fe and Mn	380	610	28	181	160	170	Valves and pump impellers
Phosphor bronze	91.5 Cu, 8 Sn, 0.5 P	376	424	65	111	190	112	Bearings and springs
Stainless steel	18 Cr, 8 Ni, 0.12 C Remainder Fe	170	460	40	195	180	260	Valves and turbine blading
Steel	0.23 C, 1 Mn, 0.5 Si, Remainder Fe	235	470	22	207	130	230	Bedplates, columns of engines and other structural items

Aluminium

Aluminium is a light material which has a good resistance to atmospheric corrosion. It is usually used as an alloy with small percentages of copper, magnesium, iron, manganese, zinc, chromium and titanium. It is also used as a minor alloying element with other metals. Suitable heat treatment can significantly improve the properties of the alloy.

Copper

Copper has good electrical conductivity and is much used in electrical equipment. It has a high resistance to corrosion and also forms a number of important alloys, such as the brasses and bronzes.

Zinc

Zinc has a good resistance to atmospheric corrosion and is used as a coating for protecting steel: the process is called 'galvanising'. It is also used as a sacrificial anode material because of its position in the galvanic series. See the subsection on corrosion later in this Chapter. A number of alloys are formed using zinc.

Tin

A ductile, malleable metal that is resistant to corrosion by air or water. It is used as a coating for steel and also in various alloys.

Titanium

A light, strong, corrosion-resistant metal which is used as the plate material in plate-type heat exchangers. It is also used as an alloying element in various special steels.

Brass

An alloy of copper and zinc with usually a major proportion of copper. A small amount of arsenic may be added to prevent a form of corrosion known as 'dezincification' occurring. Other alloying metals, such as aluminium, tin and manganese, may be added to improve the properties.

Bronze

An alloy of copper and tin with superior corrosion and wear resistance to brass. Other alloying additions, such as manganese, form manganese bronze or propeller brass. Additions of aluminium and zinc result in aluminium bronze and gunmetal respectively.

Cupro-nickel

An alloy of copper and nickel with 20 or 30% of nickel. Good strength properties combined with a resistance to corrosion by sea or river waters make this a popular alloy. Monel metal is a particular cupro-nickel alloy with small additions of iron, manganese, silicon and carbon.

White metal

Usually a tin based alloy with amounts of lead, copper and antimony. It may also be a lead based alloy with antimony. White metal has a low coefficient of friction and is used as a lining material for bearings.

Non-metallic materials

Many non-metallic materials are in general use. Their improved properties have resulted in their replacing conventional metals for many applications. The majority are organic, being produced either synthetically or from naturally occurring material.

Ceramics are being increasingly considered for marine use particularly where galvanic corrosion is a problem. Sintered alpha silicon carbide and other silicon-based ceramics have good strength properties and are inert in sea water.

The general term 'plastic' is used to describe many of these non-metallic materials. Plastics are organic materials which can be moulded to shape under the action of heat or heat and pressure. There are two main classes, thermoplastic and thermosetting, although some more modern plastics are strictly neither. Thermoplastic materials are softened by heat and can be formed to shape and then set by cooling, e.g. perspex, polyvinylchloride (PVC) and nylon. Thermosetting materials are usually moulded in a heated state, undergo a chemical change on further heating and then set hard, for example Bakelite, epoxy resins and polyesters.

Some general properties of plastic materials are good corrosion resistance, good electrical resistance and good thermal resistance; but they are unsuitable for high temperatures. To improve or alter properties, various additives or fillers are used, such as glass fibre for strength. Asbestos fibre can improve heat resistance and mica is sometimes added to reduce electrical conductivity.

Foamed plastics are formed by the liberation of gas from the actual material, which then expands to form a honeycomb-like structure. Such materials have very good sound and heat insulating properties. Many plastics can be foamed to give low density materials with a variety of properties, for example, fire extinguishing. Some well known non-metallic materials are:

Asbestos

A mineral which will withstand very high temperatures and is unaffected by steam, petrol, paraffin, fuel oils and lubricants. It is used in many forms of jointing or gasket material and in various types of gland packing. It does however present a health hazard in some forms.

Cotton

A fibrous material of natural origin which is used as a backing material for rubber in rubber insertion jointing. It is also used in some types of gland packing material.

Glass reinforced plastics (GRP)

A combination of thin fibres of glass in various forms which, when mixed with a resin, will cure (set) to produce a hard material which is strong and chemically inert. It has a variety of uses for general repairs.

Lignum vitae

A hardwood which is used for stern bearing lining. It can be lubricated by sea water but is subject to some swelling.

Nylon

A synthetic polymer which is chemically inert and resistant to erosion and impingement attack. It is used for orifice plates, valve seats and as a coating for salt water pipes.

Polytetrafluorethylene (PTFE)

A fluoropolymer which is chemically inert and resistant to heat. It has a low coefficient of friction and is widely used as a bearing material. It can be used dry and is employed in sealed bearings. Impregnated with graphite, it is used as a filling material for glands and guide rings.

Polyvinylchloride (PVC)

A vinyl plastic which is chemically inert and used in rigid form for pipework, ducts, etc. In a plasticised form it is used for sheeting, cable covering and various mouldings.

Resin

Resins are hard, brittle substances which are insoluble in water. Strictly speaking they are added to polymers prior to curing. The term 'resin' is often incorrectly used to mean any synthetic plastic. Epoxy resins are liquids which can be poured and cured at room temperatures. The cured material is unaffected by oils and sea water. It is tough, solid and durable and is used as a chocking material for engines, winches, etc.

Rubber

A tree sap which solidifies to form a rough, elastic material which is unaffected by water but is attacked by oils and steam. It is used as a jointing material for fresh and sea water pipes and also for water lubricated bearings. When combined with sulphur (vulcanized) it forms a hard material called 'ebonite' which is used for bucket rings (piston rings) in feed pumps. Synthetic rubbers such as neoprene and nitrile rubber are used where resistance to oil, mild chemicals or higher temperatures is required.

Joining metals

Many larger items of engineering equipment are the result of combining or joining together smaller, easily produced items. Various joining methods exist, ranging from mechanical devices, such as rivets or nuts and bolts, to fusion welding of the two parts.

It is not proposed to discuss riveting, which no longer has any large scale marine engineering applications, nor will nuts and bolts be mentioned, since these are well known in their various forms.

Brazing and soldering are a means of joining metal items using an alloy (solder) of lower melting point than the metals to be joined. The liquefied solder is applied to the heated joint and forms a very thin layer of metal which is alloyed to both surfaces. On cooling the two metals are joined by the alloy layer between them.

Welding is the fusion of the two metals to be joined to produce a joint which is as strong as the metal itself. It is usual to join similar metals by welding. To achieve the high temperature at which fusion can take place, the metal may be heated by a gas torch or an electric arc.

With gas welding, a torch burning oxygen and acetylene gas is used and rods of the parent plate material are melted to provide the metal for the joint.

An electric arc is produced between two metals in an electric circuit when they are separated by a short distance. The metal to be welded forms one electrode in the circuit and the welding rod the other. The electric arc produced creates a region of high temperature which melts and enables fusion of the metals to take place. A transformer is used to provide a low voltage and the current can be regulated depending upon the metal thickness. The electrode provides the filler material for the joint and is flux coated to exclude atmospheric gases from the fusion process.

Corrosion

Corrosion is the wasting of metals by chemical or electrochemical reactions with their surroundings. A knowledge of the various processes or situations in which corrosion occurs will enable at least a slowing down of the material wastage.

Iron and steel corrode in an attempt to return to their stable oxide form. This oxidising, or 'rusting' as it is called, will take place wherever steel is exposed to oxygen and moisture. Unfortunately the metal oxide formed permits the reaction to continue beneath it. Some metals however produce a passive oxide film, that is, no further corrosion takes place beneath it. Aluminium and chromium are examples of metals which form passive oxide coatings. Corrosion control of this process usually involves a coating. This may be another metal, such as tin or zinc, or the use of paints or plastic coatings.

Electrochemical corrosion usually involves two different metals with an electrolyte between them. (An electrolyte is a liquid which enables current to flow through it.) A corrosion cell or galvanic cell is said to have been set up. Current flow occurs in the cell between the two metals since they are at different potentials. As a result of the current flow through the electrolyte, metal is removed from the anode or positive electrode and the cathode or negative electrode is protected from corrosion. A corrosion cell can occur between different parts of the same metal, perhaps due to slight variations in composition, oxygen concentration, and so on. The result is usually small holes or pits and the effect is known as 'pitting corrosion'. A more serious form of this effect results in greater damage and is known as 'crevice corrosion'. The prevention of electrochemical corrosion is achieved by cathodic protection. Pitting and crevice corrosion can be countered by a suitable choice of materials, certain copper alloys for instance.

'Erosion' is a term often associated with corrosion and is the wearing away of metal by abrasion. Sea water systems are prone to problems of this nature. Increasing water velocity can reduce pitting problems in some materials but will increase their general surface corrosion, for example copper base alloys. Water impinging on a surface can cause erosion damage and this is usually found where turbulent flow conditions occur. The water inlet tube plates of heat exchangers often suffer from this problem. Careful material selection is necessary to reduce this type of erosion.

Cathodic protection

Cathodic protection operates by providing a reverse current flow to that of the corrosive system. This can be achieved in one of two ways, either by the use of sacrificial anodes or an impressed current. The sacrificial anode method uses metals such as aluminium or zinc which form the anode of the corrosion cell in preference to other metals such as steel. The impressed current system provides an electrical potential difference from the ship's power supply system through a long lasting anode of a corrosion resistant material such as platinised titanium.

The sacrificial anode method is employed in heat exchanger water boxes where a block or plate of zinc or aluminium is fastened to the cast iron or steel. A good connection is necessary to ensure the conduct of electricity. If the anode does not corrode away then it is not properly connected. The anode should not be painted or coated in any way.

The impressed current method is not normally used for the protection of machinery or auxiliary equipment handling sea water because of the problems of current and voltage control. It is however often used for protection of the external hull.

Chapter 17
Watchkeeping and equipment operation

The 'round the clock' operation of a ship at sea requires a rota system of attendance in the machinery space. This has developed into a system of watchkeeping that has endured until recently. The arrival of 'Un-attended Machinery Spaces' (UMS) has begun to erode this traditional practice of watchkeeping. The organisation of the Engineering Department, conventional watchkeeping and UMS practices will now be outlined.

The Engineering Department

The Chief Engineer is directly responsible to the Master for the satisfactory operation of all machinery and equipment. Apart from assuming all responsibility his role is mainly that of consultant and adviser. It is not usual for the Chief Engineer to keep a watch.

The Second Engineer is responsible for the practical upkeep of machinery and the manning of the engine room: he is in effect an executive officer. On some ships the Second Engineer may keep a watch.

The Third and Fourth Engineers are usually senior watchkeepers or engineers in charge of a watch. Each may have particular areas of responsibility, such as generators or boilers.

Fifth and Sixth Engineers may be referred to as such, or all below Fourth Engineer may be classed as Junior Engineers. They will make up as additional watchkeepers, day workers on maintenance work or possibly act as Refrigeration Engineer.

Electrical Engineers may be carried on large ships or where company practice dictates. Where no specialist Electrical Engineer is carried the duty will fall on one of the engineers.

Various engine room ratings will usually form part of the engine room complement. Donkeymen are usually senior ratings who attend the auxiliary boiler while the ship is in port. Otherwise they will direct the ratings in the maintenance and upkeep of the machinery space. A storekeeper may also be carried and on tankers a pump man is employed to maintain and operate the cargo pumps. The engine room ratings, e.g. firemen, greasers, etc., are usually employed on watches to assist the engineer in charge.

The watchkeeping system

The system of watches adopted on board ship is usually a four hour period of working with eight hours rest for the members of each watch. The three watches in any 12 hour period are usually 12–4, 4–8 and 8–12. The word 'watch' is taken as meaning the time period and also the personnel at work during that period.

The watchkeeping arrangements and the make up of the watch will be decided by the Chief Engineer. Factors to be taken into account in this matter will include the type of ship, the type of machinery and degree of automation, the qualifications and experience of the members of the watch, any special conditions such as weather, ship location, international and local regulations, etc. The engineer officer in charge of the watch is the Chief Engineer's representative and is responsible for the safe and efficient operation and upkeep of all machinery affecting the safety of the ship.

Operating the watch

An engineer officer in charge, with perhaps a junior engineer assisting and one or more ratings, will form the watch. Each member of the watch should be familiar with his duties and the safety and survival equipment in the machinery space. This would include a knowledge of the fire fighting equipment with respect to location and operation, being able to distinguish the different alarms and the action required, an understanding of the communications systems and how to summon help and also being aware of the escape routes from the machinery space.

At the beginning of the watch the current operational parameters and the condition of all machinery should be verified and also the log readings should correspond with those observed. The engineer officer in charge should note if there are any special orders or instructions relating to the operation of the main machinery or auxiliaries. He should determine what work is in progress and any hazards or limitations this presents. The levels of tanks containing fuel, water, slops, ballast, etc., should be noted and also the level of the various bilges. The operating mode of equipment and available standby equipment should also be noted.

At appropriate intervals inspections should be made of the main propulsion plant, auxiliary machinery and steering gear spaces. Any routine adjustments may then be made and malfunctions or breakdowns can be noted, reported and corrected. During these tours of inspection bilge levels should be noted, piping and systems observed for leaks, and local indicating instruments can be observed.

Where bilge levels are high, or the well is full, it must be pumped dry. The liquid will be pumped to an oily water separator, and only clean water is to be discharged overboard. Particular attention must be paid to the relevant oil pollution regulations both of a national and international nature, depending upon the location of the ship. Bilges should not be pumped when in port. Oily bilges are usually emptied to a slop tank from which the oil may be reclaimed or discharged into suitable facilities when in port. The discharging of oil from a ship usually results in the engineer responsible and the master being arrested.

Bridge orders must be promptly carried out and a record of any required changes in speed and direction should be kept. When under standby or manoeuvring conditions with the machinery being manually operated the control unit or console should be continuously manned.

Certain watchkeeping duties will be necessary for the continuous operation of equipment or plant—the transferring of fuel for instance. In addition to these regular tasks other repair or maintenance tasks may be required of the watchkeeping personnel. However no tasks should be set or undertaken which will interfere with the supervisory duties relating to the main machinery and associated equipment.

During the watch a log or record will be taken of the various parameters of main and auxiliary equipment. This may be a manual operation or provided automatically on modern vessels by a data logger. A typical log book page for a slow-speed diesel driven vessel is shown in Figure 17.1.

The hours and minutes columns are necessary since a ship, passing through time zones, may have watches of more or less than four hours. Fuel consumption figures are used to determine the efficiency of operation, in addition to providing a check on the available bunker quantities. Lubricating oil tank levels and consumption to some extent indicate engine oil consumption. The sump level is recorded and checked that it does not rise or fall, but a gradual fall is acceptable as the engine uses some oil during operation. If the sump level were to rise this would indicate water leakage into the oil and an investigation into the cause must be made. The engine exhaust temperatures should all read about the same to indicate an equal power production from each cylinder. The various temperature and pressure values for the cooling water and lubricating oil should be at, or near to, the manufacturer's designed values for the particular speed or fuel lever settings. Any high

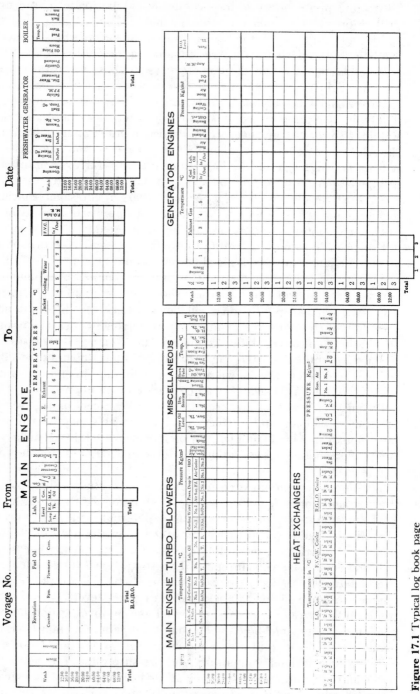

Figure 17.1 Typical log book page

outlet temperature for cooling water would indicate a lack of supply to that point.

Various parameters for the main engine turbo-blowers are also logged. Since they are high-speed turbines the correct supply of lubricating oil is essential. The machine itself is water cooled since it is circulated by hot exhaust gases. The air cooler is used to increase the charge air density to enable a large quantity of air to enter the engine cylinder. If cooling were inadequate a lesser mass of air would be supplied to the engine, resulting in a reduced power output, inefficient combustion and black smoke.

Various miscellaneous level and temperature readings are taken of heavy oil tanks, both settling and service, sterntube bearing temperature, sea water temperature, etc. The operating diesel generators will have their exhaust temperatures, cooling water and lubricating oil temperatures and pressures logged in much the same way as for the main engine. Of particular importance will be the log of running hours since this will be the basis for overhauling the machinery.

Other auxiliary machinery and equipment, such as heat exchangers, fresh water generator (evaporator), boiler, air conditioning plant and refrigeration plant will also have appropriate readings taken. There will usually be summaries or daily account tables for heavy oil, diesel oil, lubricating oil and fresh water, which will be compiled at noon. Provision is also made for remarks or important events to be noted in the log for each watch.

The completed log is used to compile a summary sheet or abstract of information which is returned to the company head office for record purposes.

The log for a medium-speed diesel driven ship would be fairly similar with probably greater numbers of cylinder readings to be taken and often more than one engine. There would also be gearbox parameters to be logged.

For a steam turbine driven vessel the main log readings will be for the boiler and the turbine. Boiler steam pressure, combustion air pressure, fuel oil temperatures, etc., will all be recorded. For the turbine the main bearing temperatures, steam pressures and temperatures, condenser vacuum, etc., must be noted. All logged values should correspond fairly closely with the design values for the equipment.

Where situations occur in the machinery space which may affect the speed, manoeuvrability, power supply or other essentials for the safe operation of the ship, the bridge should be informed as soon as possible. This notification should preferably be given before any changes are made to enable the bridge to take appropriate action.

The engineer in charge should notify the Chief Engineer in the event of any serious occurrence or a situation where he is unsure of the action to take. Examples might be, if any machinery suffers severe damage, or a malfunction occurs which may lead to serious damage. However where immediate action is necessary to ensure safety of the ship, its machinery and crew, it must be taken by the engineer in charge.

At the completion of the watch each member should hand over to his relief, ensuring that he is competent to take over and carry out his duties effectively.

UMS operation

The machinery spaces will usually be manned at least eight hours per day. During this time the engineers will be undertaking various maintenance tasks, the duty engineer having particular responsibility for the watchkeeping duties and dealing with any alarms which may occur.

When operating unmanned anyone entering the machinery space must inform the deck officer on watch. When working, or making a tour of inspection alone, the deck officer on watch should be telephoned at agreed intervals of perhaps 15 or 30 minutes.

Where the machinery space is unattended, a duty engineer will be responsible for supervision. He will normally be one of three senior watchkeeping engineers and will work on a 24 hour on, 48 hours off rota. During his rota period he will make tours of inspection about every four hours beginning at 7 or 8 o'clock in the morning.

The tour of inspection will be similar to that for a conventional watch with due consideration being given to the unattended mode of machinery operation. Trends in parameter readings must be observed, and any instability in operating conditions must be rectified, etc. A set list or mini-log of readings may have to be taken during the various tours. Between tours of inspection the Duty Engineer will be on call and should be ready to investigate any alarms relayed to his cabin or the various public rooms. The Duty Engineer should not be out of range of these alarms without appointing a relief and informing the bridge.

The main log book readings will be taken as required while on a tour of inspection. The various regular duties, such as fuel transfer, pumping of bilges, and so on, should be carried out during the daywork period, but it remains the responsibility of the Duty Engineer to ensure that they are done.

Bunkering

The loading of fuel oil into a ship's tanks from a shoreside installation or bunker barge takes place about once a trip. The penalties for oil spills are large, the damage to the environment is considerable, and the ship may well be delayed or even arrested if this job is not properly carried out.

Bunkering is traditionally the fourth engineer's job. He will usually be assisted by at least one other engineer and one or more ratings. Most ships will have a set procedure which is to be followed or some form of general instructions which might include:

1. All scuppers are to be sealed off, i.e. plugged, to prevent any minor oil spill on deck going overboard.
2. All tank air vent containments or drip trays are to be sealed or plugged.
3. Sawdust should be available at the bunkering station and various positions around the deck.
4. All fuel tank valves should be carefully checked before bunkering commences. The personnel involved should be quite familiar with the piping systems, tank valves, spill tanks and all tank-sounding equipment.
5. All valves on tanks which are not to be used should be closed or switched to the 'off' position and effectively safeguarded against opening or operation.
6. Any manual valves in the filling lines should be proved to be open for the flow of liquid.
7. Proven, reliable tank-sounding equipment must be used to regularly check the contents of each tank. It may even be necessary to 'dip' or manually sound tanks to be certain of their contents.
8. A complete set of all tank soundings must be obtained before bunkering commences.
9. A suitable means of communication must be set up between the ship and the bunkering installation before bunkering commences.
10. On-board communication between involved personnel should be by hand radio sets or some other satisfactory means.
11. Any tank that is filling should be identified in some way on the level indicator, possibly by a sign or marker reading 'FILLING'.
12. In the event of a spill, the Port Authorities should be informed as soon as possible to enable appropriate cleaning measures to be taken.

Periodic safety routines

In addition to watchkeeping and maintenance duties, various safety and emergency equipment must be periodically checked. As an example, the following inspections should take place at least weekly:

1. Emergency generator should be started and run for a reasonable period. Fuel oil, lubricating oil and cooling water supplies and tank levels should be checked.
2. Emergency fire pump should be run and the deck fire main operated for a reasonable period. All operating parameters should be checked.
3. Carbon dioxide cylinder storage room should be visually examined. The release box door should be opened to test the alarm and check that the machinery-space fans stop.
4. One smoke detector in each circuit should be tested to ensure operation and correct indication on the alarm panel. Aerosol test sprays are available to safely check some types of detector.
5. Fire pushbutton alarms should be tested, by operating a different one during each test.
6. Any machinery space ventilators or skylights should be operated and greased, if necessary, to ensure smooth, rapid closing should this be necessary.
7. Fire extinguishers should be observed in their correct location and checked to ensure they are operable.
8. Fire hoses and nozzles should likewise be observed in their correct places. The nozzles should be tried on the hose coupling. Any defective hose should be replaced.
9. Any emergency batteries, e.g. for lighting or emergency generator starting, should be examined, have the acid specific gravity checked, and be topped up, as required.
10. All lifeboat engines should be run for a reasonable period. Fuel oil and lubricating oil levels should be checked.
11. All valves and equipment operated from the fire control point should be checked for operation, where this is possible.
12. Any watertight doors should be opened and closed by hand and power. The guides should be checked to ensure that they are clear and unobstructed.

Appendix

The appendix is reserved for all those topics which, although very useful, do not quite fit in elsewhere. As an introduction, or a reminder, a section on SI units (Système International d'Unités) is given together with some conversions to the older system of Imperial units. Various engineering calculations relating to power measurement and fuel consumption are explained, together with worked examples. So that the production of manufacturing drawings of a fairly simple nature can be accomplished, an introduction to engineering drawing is provided.

SI units

The metric system of units, which is intended to provide international unification of physical measurements and quantities, is referred to as SI units.

There are three classes of units: base, supplementary and derived. There are seven base units: length—metre (m); mass—kilogram (kg); time—sound (s); electric current—ampere (A); temperature—kelvin (K); luminous intensity—candela (cd); and amount of substance—mole (mol). There are two supplementary units: plane angle—radian (rad); solid angle—steradian (sr). All remaining units used are derived from the base units. The derived units are coherent in that the multiplication or division of base units produces the derived unit. Examples of derived units are given in Table A.1.

Table A.1 Derived units

Quantity	Unit	
Force	Newton (N)	$= kg.m/s^2$
Pressure	Pascal (Pa)	$= N/m^2$
Energy, work	Joule (J)	$= N.m$
Power	Watt (W)	$= J/s$
Frequency	Hertz (Hz)	$= 1/s$

There are in use certain units which are non-SI but are retained because of their practical importance. Examples are: time—days, hours, minutes and speed—knots.

To express large quantities or values a system of prefixes is used. The use of a prefix implies a quantity multiplied by some index of 10. Some of the more common prefixes are:

$$
\begin{aligned}
1\,000\,000\,000 &= 10^9 &&= \text{giga} &&= G \\
1\,000\,000 &= 10^6 &&= \text{mega} &&= M \\
1\,000 &= 10^3 &&= \text{kilo} &&= k \\
100 &= 10^2 &&= \text{hecto} &&= h \\
10 &= 10^1 &&= \text{deca} &&= da \\
0.1 &= 10^{-1} &&= \text{deci} &&= d \\
0.01 &= 10^{-2} &&= \text{centi} &&= c \\
0.001 &= 10^{-3} &&= \text{milli} &&= m \\
0.000\,001 &= 10^{-6} &&= \text{micro} &&= \mu \\
0.000\,000\,001 &= 10^{-9} &&= \text{nano} &&= n
\end{aligned}
$$

Example:

10 000 metres = 10 kilometres = 10 km
0.001 metres = 1 millimetre = 1 mm

Note: Since kilogram is a base unit care must be taken in the use and meaning of prefixes and since only one prefix can be used then, for example, 0.000 001 kg = 1 milligram

A conversion table for some well known units is provided in Table A.2.

Table A.2 Conversion factors

To convert from	to	multiply
Length		
inch (in)	metre (m)	0.0254
foot (ft)	metre (m)	0.3048
mile	kilometre (km)	1.609
nautical mile	kilometre (km)	1.852
Volume		
cubic foot (ft^3)	cubic metre (m^3)	0.02832
gallon (gal)	litre (l)	4.546
Mass		
pound (lb)	kilogram (kg)	0.4536
tonne	kilogram (kg)	1016
Force		
pound-force (lbf)	newton (N)	4.448
ton-force	kilonewton (kN)	9.964

To convert from	to	multiply
Pressure		
pound-force per square inch (lbf/in²)	kilonewton per square metre (kN/m²)	6.895
atmosphere (atm)	kilonewton per square metre (kN/m²)	101.3
kilogram force per square centimetre (kgf/cm²)	kilonewton per square metre (kN/m²)	98.1
Energy		
foot pound-force	joule (J)	1.356
British Thermal Unit (BTU)	kilojoule (kJ)	1.055
Power		
horsepower (hp)	kilowatt (kW)	0.7457
metric horsepower	kilowatt (kW)	0.7355

Engineering terms

The system of measurement has been outlined with an introduction to SI units. Some of the common terms used in engineering measurement will now be described.

Mass

Mass is the quantity of matter in a body and is proportional to the product of volume and density. The unit is the kilogram and the abbreviation used is 'kg'. Large quantities are often expressed in tonnes (t) where 1 tonne = 10^3 kg.

Force

Acceleration or retardation of a mass results from an applied force. When unit mass is given unit acceleration then a unit of force has been applied. The unit of force is the newton (N).

force = mass × acceleration
N   kg   m/s²

Masses are attracted to the earth by a gravitational force which is the product of their mass and acceleration due to gravity (g). The value of 'g' is 9.81 m/s². The product of mass and 'g' is known as the weight of a body and for a mass 'ω' kg would be ω × g=9.81 ω newtons.

Work

When a force applied to a body causes it to move then work has been done. When unit mass is moved unit distance then a unit of work has been done. The unit of work is the joule (J).

$$\text{work} = \underset{\text{J}}{\quad} (\underset{\text{N}}{\text{mass} \times \text{g}}) \times \underset{\text{m}}{\text{distance}}$$

Power

This is the quantity of work done in a given time or the rate of doing work. When unit work is done in unit time then a unit of power has been used. The unit of power is the watt (W).

$$\text{power} = \frac{\text{mass} \times \text{g} \times \text{distance}}{\text{time}}$$

$$\text{W} = \frac{\text{N} \times \text{m}}{\text{s}}$$

Energy

This is the stored ability to do work and is measured in units of work done, i.e. joules.

Pressure

The intensity of force or force per unit area is known as pressure. A unit of pressure exists where unit force acts on unit area. The unit of pressure is the newton per square metre and has the special name pascal (Pa).

$$\text{pressure (Pa)} = \frac{\text{force}}{\text{area}} = \frac{\text{N}}{\text{m}^2}$$

Another term often used by engineers is the bar where 1 bar is equal to 10^5 Pa.

The datum or zero for pressure measurements must be carefully considered. The complete absence of pressure is a vacuum and this is therefore the absolute zero of pressure measurements. However, acting upon the earth's surface at all times is what is known as 'atmospheric pressure'. The pressure gauge, which is the usual means of pressure measurement, also accepts this atmospheric pressure and considers it the zero of pressure measurements. Thus:

absolute pressure = gauge pressure + atmospheric pressure

Readings of pressure are considered to be absolute unless followed by the word 'gauge' indicating a pressure gauge value. The actual value of atmospheric pressure is usually read from a barometer in millimetres of mercury:

atmospheric pressure = mm of mercury × 13.6 × 9.81 Pa

A standard value of 1 atmosphere is often used where the actual value is unknown,

1 atmosphere = 101300 Pa
$\qquad\qquad$ = 1.013 bar

Volume

The amount of physical space occupied by a body is called volume. The unit of volume is the cubic metre (m^3). Other units are also in use, such as litre (l) and cubic centimetre (cm^3), i.e.

$1\,m^3 = 1000\,l = 10000\,000\,cm^3$

Temperature

The degree of hotness or coldness of a body related to some zero value is known as temperature. The Celsius scale measure in °C simply relates to the freezing and boiling points of water dividing the distance shown on a thermometer into 100 equal divisions. An absolute scale has been devised based on a point 273.16 kelvin (0.01°C) which is the triple point of water. At the triple point the three phases of water can exist, i.e. ice, water and water vapour. The unit of the absolute scale is the kelvin. The unit values in the kelvin and Celsius scales are equal and the measurements of temperature are related, as follows:

$x°C = (x°C + 273)K$
or $y\,K = (y\,K - 273)°C$

Heat

Heat is energy in motion between a system and its surroundings as a consequence of a temperature difference between them. The unit, as with other forms of energy, is the joule (J).

Power measurement

The burning of fuel in an engine cylinder will result in the production of power at the output shaft. Some of the power produced in the cylinder will be used to drive the rotating masses of the engine. The power produced in the cylinder can be measured by an engine indicator mechanism as described in Chapter 2. This power is often referred to as 'indicated power'. The power output of the engine is known as 'shaft' or 'brake power'. On smaller engines it could be measured by applying a type of brake to the shaft, hence the name.

Indicated power

Typical indicator diagrams for a two-stroke and four-stroke engine are shown in Figure A.1. The area within the diagram represents the work

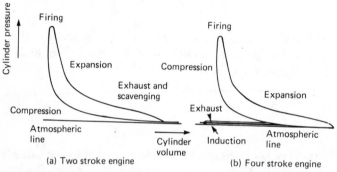

Figure A.1 Indicator diagrams

done within the measured cylinder in one cycle. The area can be measured by an instrument known as a 'planimeter' or by the use of the mid-ordinate rule. The area is then divided by the length of the diagram in order to obtain a mean height. This mean height, when multiplied by the spring scale of the indicator mechanism, gives the indicated mean effective pressure for the cylinder. The mean effective or 'average' pressure can now be used to determine the work done in the cylinder.

Work done in = mean effective × area of piston ×   length of
 1 cycle   pressure   (A)   piston stroke
                         (Pm)   (L)

To obtain a measure of power it is necessary to determine the rate of doing work, i.e. multiply by the number of power strokes in one second. For a four-stroke-cycle engine this will be rev/sec ÷ 2 and for a two-stroke-cycle engine simply rev/sec.

power developed in one cylinder

$$= \frac{\text{mean effective}}{\text{pressure (Pm)}} \times \frac{\text{area of}}{\text{piston (A)}} \times \frac{\text{length of piston}}{\text{stroke (L)}} \times \frac{\text{no. of power}}{\text{strokes/sec (N)}}$$

$$= \text{Pm L A N}$$

For a multi-cylinder engine it would be necessary to multiply by the number of cylinders.

Example

An indicator diagram taken from a six cylinder, two-stroke engine is shown in Figure A.2. The spring constant for the indicator mechanism is

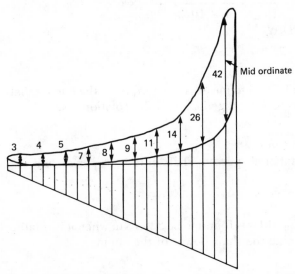

Figure A.2 Indicator diagram

65 kPa mm. The engine stroke and bore are 1100 mm and 410 mm respectively and it operates at 120 rev/min. What is the indicated power of the engine?

The diagram is divided into 10 equal parts and within each a mid-ordinate is positioned.

$$\therefore \text{ mean height of diagram} = \frac{\text{sum of mid-ordinates}}{\text{number of parts in diagram}}$$

$$= \frac{3+4+5+7+8+9+11+14+26+42}{10}$$

$$= \frac{129}{10}$$

$$= 12.9 \, \text{mm}$$

mean effective pressure, Pm = mean height of × spring
diagram   constant
= 12.9 × 65
= 838.5 kPa

$$
\begin{array}{ccccccc}
\text{indicated} & \text{mean} & \text{area} & \text{length} & \text{number of} & & \text{number} \\
\text{power of} = & \text{effective} \times & \text{of} \times & \text{of} \times & \text{power} \times & & \text{of} \\
\text{engine} & \text{pressure} & \text{piston} & \text{stroke} & \text{strokes/sec} & & \text{cylinders} \\
& (Pm) & (A) & (L) & (N) &
\end{array}
$$

$$= Pm\ L\ A\ N \times \text{number of cylinders}$$

$$= 838.5 \times \frac{1100}{10^3} \times \frac{410^2 \times \pi}{4 \times 10^6} \times \frac{120}{60} \times 6$$

$$= 1461.28\,\text{kW}$$

Shaft power

A torsionmeter is usually used to measure the torque on the engine shaft (see Chapter 15). This torque, together with the rotational speed, will give the shaft power of the engine.

shaft power = torque in shaft
$\quad\quad$ × rotational speed of shaft in radians per second

Example

The torque on an engine shaft is found to be 320 kNm when it is rotating at 110 rev/min. Determine the shaft power of the engine.

shaft power = shaft torque
$\quad\quad$ × 2π
$\quad\quad$ × revolutions per second

$$= 320 \times 2\pi \times \frac{110}{60}$$

$$= 3686.14\,\text{kW}$$

Mechanical efficiency

The power lost as a result of friction between the moving parts of the engine results in the difference between shaft and indicated power. The ratio of shaft power to indicated power for an engine is known as the 'mechanical efficiency'.

$$\text{mechanical efficiency} = \frac{\text{shaft power}}{\text{indicated power}}$$

Power utilisation

The engine shaft power is transmitted to the propeller with only minor transmission losses. The operation of the propeller results in a forward thrust on the thrust block and the propulsion of the ship at some particular speed. The propeller efficiency is a measure of effectiveness of the power conversion by the propeller.

$$\text{propeller efficiency} = \frac{\text{thrust force} \times \text{ship speed}}{\text{shaft power}}$$

The power conversion achieved by the propeller is a result of its rotating action and the geometry of the blades. The principal geometrical feature is the pitch. This is the distance that a blade would move forward in one revolution if it did not slip with respect to the water. The pitch will vary at various points along the blade out to its tip but an average value is used in calculations. The slip of the propeller is measured as a ratio or percentage as follows:

$$\text{propeller slip} = \frac{\substack{\text{theoretical speed or distance moved}\\ - \text{actual speed or distance moved}}}{\text{theoretical speed or distance moved}}$$

The theoretical speed is a product of pitch and the number of revolutions turned in a unit time. The actual speed is the ship speed. It is possible to have a negative value of slip if, for example, a strong current or wind were assisting the ship's forward motion.

Example

A ship on a voyage between two ports travels 2400 nautical miles in eight days. On the voyage the engine has made 820 000 revolutions. The propeller pitch is 6 m. Calculate the percentage propeller slip.

$$\text{theoretical distance} = \frac{820\,000 \times 6}{1852} \text{ (1 nautical mile} = 1852 \text{ metres)}$$

$$= 2656.59 \text{ n.miles}$$

$$\text{percentage slip} = \frac{\text{theoretical distance} - \text{actual distance} \times 100}{\text{theoretical distance}}$$

$$= \frac{2656.59 - 2400 \times 100}{2656.59}$$

$$= 9.66\%$$

Power estimation

The power developed by a ship's machinery is used to overcome the ship's resistance and propel it at some speed. The power required to propel a ship of a known displacement at some speed can be approximately determined using the Admiralty coefficient method.

The total resistance of a ship, R_t can be expressed as follows:

Total resistance $R_t = \rho S\, V^n$

where ρ is density (kg/m^3)
             S is wetted surface area (m^2)
             V is speed (knots)
now   Wetted surface area $\propto$ (Length)^2
             Displacement, $\Delta \propto$ (Length)^3
thus   Wetted surface area $\propto$ (Displacement, Δ)$^{2/3}$

Most merchant ships will be slow or medium speed and the index 'n' may therefore be taken as 2. The density, ρ, is considered as a constant term since all ships will be in sea water.

Total resistance, $R_t = \Delta^{2/3}V^2$
Propeller power $\propto R_t \times V$
                               $\propto \Delta^{2/3}V^2\, V$
                               $\propto \Delta^{2/3}V^3$

or   Constant $= \dfrac{\Delta^{2/3}V^3}{P}$

This constant is known as the 'Admiralty coefficient'.

Example

A ship of 15 000 tonnes displacement has a speed of 14 knots. If the Admiralty Coefficient is 410, calculate the power developed by the machinery.

Admiralty coefficient $= \dfrac{\Delta^{2/3}V^3}{P}$

Power developed, P $= \dfrac{\Delta^{2/3}V^3}{C}$

$= \dfrac{(15\,000)^{2/3}\,(14)^3}{410}$

$= 4070\,\text{kW}$

Fuel estimation

The fuel consumption of an engine depends upon the power developed. The power estimation method described previously can therefore be modified to provide fuel consumption values. The rate of fuel consumption is the amount of fuel used in a unit time, e.g. tonne/day. The specific fuel consumption is the amount of fuel used in unit time to produce unit power, e.g. kg/kW hr.

since fuel consumption $\propto$ power

where power $= \dfrac{\Delta^{2/3}V^3}{\text{Admiralty coefficient}}$

then fuel consumption/day $= \dfrac{\Delta^{2/3}V^3}{\text{fuel coefficient}}$

or fuel coefficient $= \dfrac{\Delta^{2/3}V^3}{\text{fuel consumption/day}}$

Where the fuel coefficient is considered constant a number of relationships can be built up to deal with changes in ship speed, displacement, etc.

i.e. fuel coefficient $= \dfrac{\Delta_1^{2/3}\,V_1^{\,3}}{\text{fuel cons.}_1} = \dfrac{\Delta_1^{2/3}\,V_2^{\,3}}{\text{fuel cons.}_2}$

Where 1 and 2 relate to different conditions.

$\therefore \quad \dfrac{\text{fuel cons.}_1}{\text{fuel cons.}_2} = \left(\dfrac{\Delta_1}{\Delta_2}\right)^{2/3} \times \left(\dfrac{V_1}{V_2}\right)^3$

Considering the situation on a particular voyage or over some known distance.

fuel cons./day $\times$ number of days $=$ voyage or distance consumption

or fuel cons./day $\times \dfrac{\text{voyage distance}}{\text{speed} \times 24} =$ voyage or distance consumption

Where conditions vary on different voyages or over particular distances, then:

$\dfrac{\text{voyage cons.}_1}{\text{voyage cons.}_2} = \dfrac{\text{fuel cons.}_1}{\text{fuel cons.}_2} \times \dfrac{\text{voyage distance}_1}{\text{voyage distance}_2} \times \dfrac{\text{speed}_2}{\text{speed}_1}$

and from an earlier expression

$\dfrac{\text{fuel cons.}_1}{\text{fuel cons.}_2} = \left(\dfrac{\Delta_1}{\Delta_2}\right)^{2/3} \times \left(\dfrac{V_1}{V_2}\right)^3$

$\therefore \quad \dfrac{\text{voyage cons.}_1}{\text{voyage cons.}_2} = \left(\dfrac{\text{displacement}_1}{\text{displacement}_2}\right)^{2/3} \times \left(\dfrac{\text{speed}_1}{\text{speed}_2}\right)^3 \times \left(\dfrac{\text{speed}_2}{\text{speed}_1}\right)$

$\times \dfrac{\text{voyage distance}_1}{\text{voyage distance}_2}$

This provides the general expression

$$\frac{\text{voyage cons.}_1}{\text{voyage cons.}_2} = \left(\frac{\text{displacement}_1}{\text{displacement}_2}\right)^{2/3} \times \left(\frac{\text{speed}_1}{\text{speed}_2}\right)^2 \times \frac{\text{voyage distance}_1}{\text{voyage distance}_2}$$

Example

A vessel with a displacement of 12 250 tonnes burns 290 tonnes of fuel when travelling at a speed of 15 knots on a voyage of 2850 nautical miles. For a voyage of 1800 nautical miles at a speed of 13 knots and a displacement of 14 200 tonnes estimate the quantity of fuel that will be burnt.

$$\frac{\text{voyage cons.}_1}{\text{voyage cons.}_2} = \left(\frac{\text{displacement}_1}{\text{displacement}_2}\right)^{2/3} \times \left(\frac{\text{speed}_1}{\text{speed}_2}\right)^2 \times \frac{\text{voyage distance}_1}{\text{voyage distance}_2}$$

$$\frac{\text{voyage cons.}_1}{290} = \left(\frac{14200}{12250}\right)^{2/3} \times \left(\frac{13}{15}\right)^2 \times \frac{1800}{2850}$$

$$\text{voyage cons.}_1 = 1.103 \times 0.75 \times 0.63 \times 290$$
$$= 151.14 \text{ tonnes}$$

Engineering drawing

Most engineering items defy description in words alone. To effectively communicate details of engineering equipment a drawing is usually used. Even the simplest of sketches must conform to certain rules or standards to ensure a 'language' that can be readily understood.

Some of these basic rules will now be described with the intention of enabling the production of a simple drawing for manufacturing or explanation purposes. A drawing produced as a piece of information or communication should stand alone, that is, no further explanation should be necessary. All necessary dimensions should be provided on the drawing and the materials to be used should be specified.

A drawing is made up of different types of lines, as shown in Figure A.3. The continuous thick line is used for outlining the drawing. The continuous thin line is used for dimension lines, to indicate sectioning, etc. A series of short dashes represents a hidden detail or edge and a chain dotted line is used for centre lines.

To represent a three dimensional item in two dimensions a means of projecting the different views is necessary. Two systems of projection are in use, *First Angle* and *Third Angle*. The First Angle system will be described with reference to the object shown in Figure A.4. Three views are drawn by looking at the object in the directions 1, 2 and 3. The views

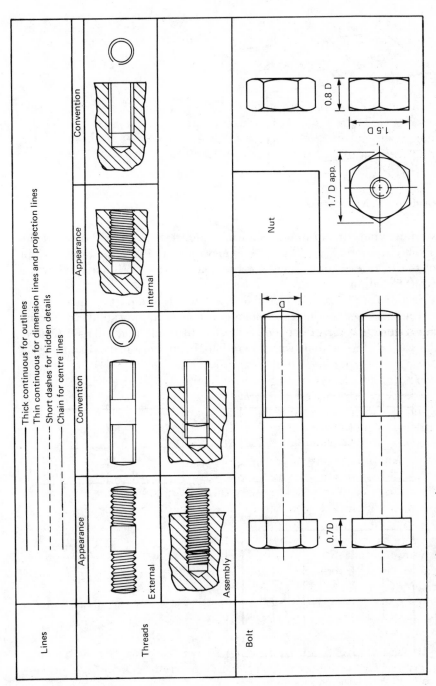

Figure A.3 Conventional representation

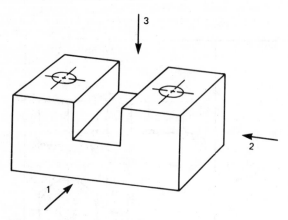

Figure A.4 Views for projection

seen are then drawn out, as shown in Figure A.5. View 1 is called the 'front elevation'. View 2 is the 'end elevation' and is located to the right of the front elevation. View 3 is the 'plan' and is positioned below the front elevation.

Sections are used to show the internal details of a part or an assembly as full lines. Section lines or hatching are used to indicate the different items which have in effect been cut. Each different item will have section lines at a different angle with 45° and 60° being most usual. Examples of

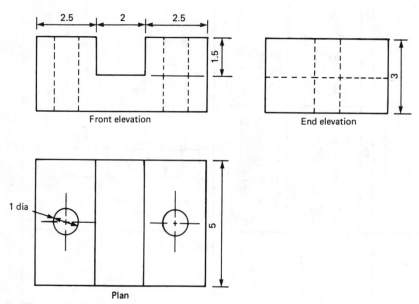

Figure A.5 First angle projection

a sectioning can be seen for the internal threads shown in Figure A.3. The plane of section is always given and is shown on other views of the item.

Dimension lines are essential for the manufacture of an item. The dimension line is a thin continuous line with an arrowhead at each end and the dimension is placed above it at right angles. Where possible projection lines are used to allow the dimension line to be clear of the drawing. The projection line begins a small distance clear of the drawing outline. Leader lines are used to indicate information to the appropriate part of the drawing and an arrowhead is used at the end of the line.

Scales are used to reduce drawings to reasonable sizes while retaining the correct proportions. Standard scale reductions are 1:1; 1:2; 1:5; 1:10 etc., where for example 1:10 means one-tenth full size. The in-between scale sizes are not normally used. Special scale rules are available to simplify drawing in any one of the above scales.

Standard representations are used for common engineering items such as nuts, bolts, studs, internal and external threads, etc. These are shown in Figure A.3. The proportions used for drawing nuts and bolts should be remembered and used whenever necessary.

A final word on the subject of information is necessary. A drawing should enable the item to be manufactured or at least identified so that a replacement can be obtained. Apart from the actual drawing there should be a block of information giving the item name, any material to be used, the drawing scale, stating the projection and possibly the date and the name of the person who made the drawing.

Index

Absolute
  pressure, 278, 350, 351
  temperature, 351
Acid and basic processes, 329
Acidity, 152
Actuator, 309
Admiralty coefficient, 356
Aerobic bacteria, 147
Air compressors, 134–137
  automatic valves, 137
  water jacket safety valve, 135
Air conditioning, 163, 175–177
  single duct, 175
  single duct with reheat, 175, 177
  twin duct, 175–177
Air ejector, 105, 106
Air register, 91, 92
Air release cock, 85
Air supply, 91
ALCAP, 157
Alkalinity, 94
Alternating current
  distribution, 261, 262
  generators, 258–261
  motors, 266–269
  paralleling of generators, 263
  positioning motors, 300
  supply, 263
Alternator
  brushless high-speed, 261
  shaft-driven, 261
  statically excited, 260
Aluminium, 333
Ammeter, 257, 288
Amplifier, 300
Anchor capstan, 183
Annealing, 330
Asbestos, 335
Astern turbine, 57
Atmospheric drain tank, 101, 102
Atmospheric pressure, 278, 350, 351
Atomisation, 28
Attemperator, 74
Auto pilot, 321–323

Auto transformer starting, 267, 268
Automatic
  feed water regulator, 85
  self-tensioning winches, 181
  voltage regulator, 260
  water spray, 239, 240
Auxiliary steam
  plant, 84
  stop valve, 85
Axial thrust, 59, 60

Ballast system, 132
Barometer
  aneroid, 280
  mercury, 279
Batteries, 269–273
  alkaline, 270, 271
  charging, 271
  lead–acid, 270
  maintenance, 272, 273
  operating characteristics, 271, 273
  selection, 271
Bearings
  slow-speed diesel, 29
  steam turbine, 62, 63
Bedplate, 44, 47
Bellmouth, 129
Bellows pressure gauge, 281
Bend test, 328
Bilge
  injection valve, 131
  system, 131
Blades
  impulse, 54
  reaction, 54, 55
  root fastening, 58, 59
Blow down valve, 85
Boiler
  blowbacks, 251
  combustion control, 313, 314
  composite, 82, 85
  D-type, 76
  ESD-type, 77
  mountings, 85–91

363